The Trans-Atlantic Pioneers

The Trans-Atlantic Pioneers

From First Flights to Supersonic Jets – The Battle to Cross the Atlantic

Bruce Hales-Dutton

First published in Great Britain in 2018 by
AIR WORLD
An imprint of
Pen & Sword Books Ltd
Yorkshire - Philadelphia

ISBN 978 1 52673 2 170

A CIP catalogue record for this book is
available from the British Library

Typeset in 10.5/13.5 Ehrhardt MT by SRJ Info Jnana System Pvt Ltd.
Printed and bound by TJ International

Pen & Sword Books Ltd incorporates the imprints of Pen & Sword Archaeology, Atlas, Aviation, Battleground, Discovery, Family History, History, Maritime, Military, Naval, Politics, Social History, Transport, True Crime, Claymore Press, Frontline Books, Praetorian Press, Seaforth Publishing and White Owl

For a complete list of Pen & Sword titles please contact

PEN & SWORD BOOKS LTD
47 Church Street, Barnsley, South Yorkshire, S70 2AS, England
E-mail: enquiries@pen-and-sword.co.uk
Website: www.pen-and-sword.co.uk

Or

PEN AND SWORD BOOKS
1950 Lawrence Rd, Havertown, PA 19083, USA
E-mail: Uspen-and-sword@casematepublishers.com
Website: www.penandswordbooks.com

Contents

Introduction

In the summer of 1919, barely six months after the Armistice, Britain was able to acclaim its first peacetime heroes.

Winston Churchill was probably speaking for the nation when he compared the achievement of John Alcock and Arthur Whitten Brown with that of Christopher Columbus four centuries earlier.

At a lunch to honour the two aviators, and to hand them the £10,000 *Daily Mail* prize for making the first non-stop flight across the Atlantic Ocean, Churchill also evoked visions of the closer ties between Europe and North America that would surely result.

The Secretary of State for War also drew attention to the rapid strides made by aeronautical technology when he pointed out that the English Channel had been crossed by air for the first time only a decade earlier. He might have added that by 1919 powered flight was scarcely sixteen years old.

But anybody who expected to be buying a ticket to fly from London to New York would be waiting rather a long time. Two decades after Alcock and Brown's achievement aircraft could fly at 400mph and reach altitudes of 10 miles, but scheduled commercial trans-Atlantic travel still lay sometime in the future; and even longer for those without deep pockets.

It would be four decades after that lunch at London's posh Savoy Hotel that more people were crossing the Atlantic by air than by sea. The march of technology continued, however, and another decade and a half further on it was possible for 100 passengers to cross the ocean while flying on the fringe of space at the speed of a rifle bullet.

Advancing technology has been one of the keys to aeronautical progress, but back in 1919 Churchill also talked about the 'peril and the pluck' that had characterised Alcock and Brown's achievement. He declared: 'I really do not know what we should admire the most in our guests – their audacity, their determination, their skill, their science, their Vickers Vimy aeroplane, their Rolls-Royce engines or their good fortune.'

And, indeed, each step towards grasping the complex skills needed to fly the unforgiving and ever-changing Atlantic Ocean in ever-increasing safety and comfort has required a blend of technology with human skill and guts.

As a story of human achievement it is rich in characters. It starts with newspaper magnate Lord Northcliffe and his £10,000 prize; Albert C. Reade, the laconic commander of the first aircraft to make the crossing; the fearless Harry Hawker, who made failure seem like success; the boyish John Alcock and the studious Arthur Whitten Brown, whose fierce desire to fly the Atlantic sustained them through wartime captivity; Jack Pritchard who, by parachuting from the British airship *R34* to give mooring instructions, became the first trans-Atlantic aviator to set foot on US soil; the lanky Charles Lindbergh, whose triumph ignited the USA's enthusiasm for aviation; and the tomboyish Amelia Earhart, who became the first woman pilot to make the crossing.

There were plenty more: Percy Maxwell Muller, who helped convince sceptical Vickers directors that the company should compete for the *Daily Mail* Atlantic prize; 'Rex' Pierson, who designed the Vimy that first demonstrated the peaceful potential of long-distance aviation; Hugo Eckener, the airship devotee whose attitude so upset his Nazi bosses; Juan Trippe, whose vision of mass travel led Pan American to operate the first commercial trans-Atlantic services; Donald Bennett, the straight-talking Imperial Airways captain and navigator who commanded the first ferry flight of much-needed warplanes in November 1940; the spectacularly eccentric Howard Hughes, who cajoled and bullied Lockheed into making its Constellation the first truly capable trans-Atlantic airliner; Geoffrey de Havilland, whose foresight and determination brought scheduled jets to the North Atlantic; Bill Allen, the Boeing president who persuaded his board to commit the funds to launch the 747; and to Freddie Laker, whose persistence in the face of official hostility led to the low-cost Skytrain service. There are many others and no doubt there will be more in the next 100 years.

All brought their own special qualities and talents to the proposition that crossing the Atlantic by air was possible and practical. In their different ways they were trans-Atlantic pioneers. Their legacy is the daily flow across the North Atlantic of up to 3,000 aircraft flying on more than 400 different routes between airports in Europe and North America.

I hope this book will act as a tribute to them all.

Bruce Hales-Dutton,
West Malling, Kent,
March 2018.

Trans-Atlantic Chronology

1859

John Wise announces plans to fly the Atlantic in a balloon, but he crash-lands in New York state.

1910

15 October	Walter Wellman makes the first attempt to cross the Atlantic in the dirigible *America* but he is forced to ditch in the sea after about 1,000 miles.

1913

1 April	The *Daily Mail* announces a prize of £10,000 for the first to fly non-stop across the Atlantic.

1918

November	The Royal Aero Club publishes rules for revived *Daily Mail* Atlantic contest.

1919

April–June	Contestants for the trans-Atlantic attempt assemble at St John's, Newfoundland.
8 May	US Navy flying boats NC-1, NC-3 and NC-4 takeoff from Jamaica Bay for Halifax on first stage of attempt to fly the Atlantic by aeroplane.
16 May	US Navy flying boats leave Trepassey harbour, Newfoundland.
18 May	Harry Hawker and Kenneth Mackenzie Grieve take off on the first attempt to fly the Atlantic non-stop and claim the *Daily Mail* prize but the aviators are forced to ditch in the sea early the following day to be rescued by a Danish steamer; Raynham and Morgan crash on take-off in Martinsyde Raymor; Raynham subsequently tried

	again with a different navigator but the aircraft crashed on take-off and was badly damaged.
20 May	NC-4 commanded by Lieutenant Commander Albert C. Read reaches the Azores.
27 May	NC-4 arrives in Lisbon.
25 May	Hawker and Mackenzie Grieve arrive in Scotland; they reach London two days later.
26 May	SS *Glendevon* arrives at St John's with Vickers Vimy.
31 May	NC-4 lands at Plymouth at conclusion of first trans-Atlantic flight.
14–15 June	Alcock and Brown complete first non-stop trans-Atlantic flight from St John's to Clifden, Ireland.
6 July	His Majesty's Airship *R34* arrives at Mineola, Long Island, New York, at the conclusion of the first Atlantic flight from east to west.
13 July	*R34* arrives back at Pulham, Norfolk, to complete first double crossing of the Atlantic by air.

1922

30 March–17 June	First aerial crossing of South Atlantic by Portuguese naval aviators Gago Coutinho and Sacadura Cabral, who flew from Lisbon to Rio de Janeiro (5,209 miles/8,383km).

1924

8 April–28 September	US Army completes World Cruise.
12–15 October	Zeppelin *LZ-126* makes non-stop trans-Atlantic crossing from Friedrichshafen to Lakehurst, New Jersey.

1926

20 September	Rene Fonck crashes on take-off while making first attempt to fly east–west trans-Atlantic crossing and first non-stop flight from Paris to New York.

1927

8 May	Charles Nungesser and Francois Coli leave Paris on flight to New York but disappear, never to be seen again.
20–21 May	Charles Lindbergh completes first non-stop flight from New York to Paris.
4–6 June	Charles A. Levine becomes first trans-Atlantic passenger when he flies from New York to Eisleben, Germany, in a Bellanca piloted by Clarence Chamberlin.

16 June	General Francesco de Pinedo, with his crew of Captain Carlo del Prete and mechanic Vitale Zacchetti, arrives in Rome to complete epic 'Four Continents' flight after a journey of 29,180 miles (46,960km) taking 124 days and covering South America, the Caribbean, USA and Canada in two Savoia-Marchetti S.55 flying boats (the first was wrecked); flight home from Newfoundland via the Azores follows 1919 route of US Navy NC-4; receives the British Air Force Cross, as well as the US Distinguished Flying Cross by special Act of Congress.
1 July	Richard E. Byrd and Floyd Bennett land in Normandy after failing to find Paris on flight from New York.
14–15 October	Dieudonne Costes and Joseph le Brix make first non-stop aerial crossing of the South Atlantic flying a Breguet 19.

1928

12–13 April	Freiherr von Hunefeld, Herman Kohl and James Fitzmaurice complete first east–west trans-Atlantic crossing from Ireland to New York in Junkers W33 monoplane *Bremen*.
18 June	Amelia Earhart becomes first woman to fly the Atlantic as a passenger.
11–15 July	*Graf Zeppelin* makes first commercial trans-Atlantic flight from Friedrichshafen, Germany to Lakehurst, New Jersey.

1929

16 July	The German liner *Bremen* completes maiden Atlantic voyage, which marked the first time mail was carried by a ship-launched floatplane; a Heinkel He 12 piloted by Baron Jobst von Studnitz was launched 20 miles (32km) from US coast. It was carrying 11,000 pieces of mail in six bags, weighing 220lb (100kg), which it delivered to New York many hours before the *Bremen* docked.
7 August	*Graf Zeppelin* makes round the world flight.

1930

1–2 April	Yancey, Alexander and Bouk make first flight from New York to Bermuda.

1–2 September	Dieudonne Costes and Maurice Bellonte make the first non-stop east–west aeroplane flight between Europe and America, flying from Paris to New York.

1931

15 January	Marshal Italo Balbo returns to Rome after first formation crossing of the South Atlantic to Rio de Janeiro by twelve Savoia-Marchetti S.55 flying boats.
23 June–1 July	Wiley Post and Harold Gatty complete first round-the-world flight.
Summer	*Graf Zeppelin* starts first regular trans-Atlantic passenger flights between Germany and South America. Between then and 1937, this airship crossed the Atlantic 136 times.
27–28 November	Bert Hinkler makes first solo crossing of the South Atlantic, flying de Havilland Puss Moth from Canada to USA, then to West Indies, Venezuela, Guiana, and finally Brazil to Britain.

1932

20–21 May	Amelia Earhart makes first solo trans-Atlantic flight by a female pilot.
18 August	Jim Mollison makes first solo east–west trans-Atlantic flight, from Portmarnock, Ireland to Pennfield, New Brunswick.

1933

6–9 February	Jim Mollison flies a Puss Moth from Senegal to Brazil, across the South Atlantic, becoming the first person to fly solo across the North and South Atlantics.
1 July–12: August	Marshal Italo Balbo's Italian Air Armada of twenty-four Savoia-Marchetti S.55 flying boats make first formation crossing of the North Atlantic, flying from Rome to Chicago and back.
22 July	Wiley Post returns to New York having completed first solo round-the-world flight in seven days, eighteen hours and forty-nine minutes.

1934

February	Deutsche Luft Hansa begins regular airmail service between Natal, Brazil and Bathurst, Gambia; will continue until August 1939.

1935

December	Air France launches weekly airmail service between Africa and South America.

1936

6–9 May	*Hindenburg* begins regular passenger flights between Germany and USA.
9 October	*Hindenburg* leaves Lakehurst to begin its last eastward trip.

1937

6 May	*Hindenburg* destroyed by fire at Lakehurst, New Jersey, ending commercial trans-Atlantic airship operations.
16 June	Pan American and Imperial Airways launch New York–Bermuda flying boat services.
18–20 June	Valery Chkalov make first trans-Polar flight from Moscow to Vancouver.
5 July	An Imperial Airways Short Empire flying boat makes first of a series of proving flights between Foynes, Ireland, to Botwood, Newfoundland; Pan American Sikorsky S-42 makes simultaneous flight in opposite direction.

1938

18 July	Douglas 'Wrong Way' Corrigan lands in Ireland after 'mistaken' flight from New York.
21 July	The floatplane *Mercury*, upper component of Short-Mayo composite, completes first trans-Atlantic crossing.
9 July	Pan American Boeing 314 flying boat *Yankee Clipper* completes the first commercial trans-Atlantic flying boat service.
10 August	Deutsche Luft Hansa Focke-Wulf Fw 200 Condor makes first non-stop flight from Berlin to York; it takes twenty-four hours fifty-six minutes. It completes the return flight three days later in nineteen hours forty-seven minutes.

1940

10 November	First US-built Lockheed Hudsons leave Newfoundland on the first ferry flight to the UK.

1941

January	First US-built warplanes leave Bermuda on delivery flight to UK.
July	The Atlantic Ferry Organisation becomes part of RAF Ferry Command; Return Ferry Service begins operations.

1943

Ferry Command becomes part of RAF Transport Command

1944

1 June	US Navy completes first trans-Atlantic crossing by non-rigid airships or blimps.
November	Fifty-two nations sign the Chicago Convention, which will act as framework for post-war commercial aviation.

1945

24 October	American Overseas Airlines begins scheduled passenger service between New York and Bournemouth.

1946

5 February	Trans World Airways introduces Lockheed Constellation to North Atlantic with launch of scheduled services between New York and Paris.
1 July	British Overseas Airways Corporation launches regular London–New York Constellation service.

1948

July	De Havilland Vampires of 54 Squadron RAF become first jet aircraft to fly the Atlantic.

1949

3 June	Pan American introduces the Boeing Stratocruiser on North Atlantic route with luxurious President service that sets new standard of passenger travel.

1950

1 July	The Douglas DC-6 introduced on Rome–New York route.

1951

21 February	English Electric Canberra becomes first jet aircraft to make non-stop Atlantic crossing.

1952

December	Start of Operation *Becher's Brook*, in which more than 400 Canadian-built Sabre jet fighters are ferried to the UK; completed May 1954.

1954

15 November	Scandinavian Airlines System launches scheduled trans-polar flights between Copenhagen and Los Angeles using Douglas DC-6Bs.

1957

16 December	BOAC launches first commercial trans-Atlantic service by turbine-powered aircraft using the Bristol Britannia.
24 December	A Britannia operated by the Israeli airline El Al makes the fastest ever trans-Atlantic crossing by commercial airliner.

1958

Passengers flying across the Atlantic exceed those travelling by sea for the first time.

4 October	BOAC launches the world's first commercial jet service using de Havilland Comet 4s.
26 October	Pan American launches Boeing 707 trans-Atlantic operations by inaugurating New York–Paris service.

1963

Caledonian Airways receives foreign carrier's permit enabling it to operate charter services to USA; by 1970 it will become the leading operator of controversial 'affinity group charters'.

1969

First flights of Boeing 747 and Anglo–French supersonic Concorde.

3 June	Replica Vickers Vimy makes first flight.

1970

22 January	Pan American completes the first scheduled service by a Boeing 747 'Jumbo Jet' when *Clipper Young American* arrives at Heathrow from New York.

1971

September	The Anglo–French Concorde 001 supersonic airliner makes first Atlantic crossing, from Toulouse to Rio de Janeiro, to start two-week tour of South America.

1973

26 September	Concorde 02 makes first flight over North Atlantic by supersonic airliner, reaching Paris three hours thirty-three minutes after leaving Washington.

1974

17 June	An Air France Concorde makes first double Atlantic crossing by supersonic airliner.
September	Concorde 02 makes tour of US Pacific coast.
1 September	A USAF Lockheed SR-71 reconnaissance aircraft makes the fastest-ever Atlantic crossing in a time of one hour fifty-four minutes and 56.4 seconds. Flown by the USAF crew of James V. Sullivan (pilot) and Noel F. Widdifield, (reconnaissance systems officer), the SR-71 achieves an average velocity of about Mach 2.72, including deceleration for in-flight refuelling. Peak speeds during this flight are likely closer to the declassified top speed of more than Mach 3.2.

1975

4 October	Concorde 02 makes first supersonic airliner flight to Canada, flying to Montreal from Paris via London and Ottawa.

1976

21 January	Air France launches first supersonic commercial airliner services between Europe and American continent with start of services between Paris and Rio de Janeiro; British Airways launches parallel service between London and Bahrain.

24 May	BA and Air France launch first commercial supersonic trans-Atlantic services from London and Paris to Washington.

1977

21 May	Concorde flies from New York to Paris to mark fiftieth anniversary of Charles Lindbergh's flight.
July	Britain and USA agree new Bermuda II air services agreement, which will lead to new air services between the two countries.
September	Ben Abruzzo and Maxie Anderson ditch in the sea near Iceland after their first attempt to cross the Atlantic by balloon ends in failure after sixty-six hours and 2,950 miles (4,720km); since 1859 there have been seventeen unsuccessful attempts to cross the Atlantic by balloon with at least seven fatalities.
26 September	Laker Airways launches first low-cost trans-Atlantic services with start of Skytrain operation between London Gatwick and New York's JFK International.
22 November	BA and Air France begin Concorde services to JFK.

1978

17 August	*Double Eagle II* completes first trans-Atlantic balloon flight when Ben Abruzzo, Maxie Anderson and Larry Newman complete six-day 3,233-mile (5,173km) journey from Maine to Paris.

1979

12 January	Braniff International inaugurates subsonic Concorde service between Washington and Dallas/Fort Worth.
16 December	BA Concorde makes fastest-ever Atlantic crossing by an airliner, flying from London to New York in two hours fifty-nine minutes.

1983

May 26	People Express launches low-cost flights between JFK and Gatwick.

1984

April	El Al operates first commercial trans-Atlantic operation by twin-engine Boeing 767 under sixty-minute ETOPs rules.

18 July — Captain Lynn Rippelmeyer of People Express becomes first female pilot to command a trans-Atlantic scheduled flight.

1985

1 February — A TWA Boeing 767 makes the first approved 120-minute ETOPs flight, from Boston to Paris.

1991

Number of passengers crossing the Atlantic on twin-engine airliners exceeds that travelling on three- and four-engine jets.

1993

25 March — BA first officer Barbara Harmer becomes first female Concorde pilot to operate a scheduled trans-Atlantic service.

2000

25 July — Air France Flight 4590, registration F-BTSC, crashes after departing from Paris Charles de Gaulle en route to JFK, killing all 100 passengers and nine crew members on board, and four people on the ground. It is the only fatal accident involving Concorde.

2002

2 May — Eric Lindbergh, grandson of Charles Lindbergh, celebrates the seventy-fifth anniversary of the *Spirit of St. Louis*' 1927 flight by flying the same route in a single-engine, two-seat Lancair Columbia 200; he takes seventeen hours seven minutes, little more than half the time taken by his grandfather in 1927.

2003

BA and Air France formally withdraw their Concordes from service.

2005

2/3 July — Steve Fossett and Mark Rebholz make trans-Atlantic flight in second replica Vickers Vimy.

2006

26 August — Steve Fossett and Mark Rebholz's replica Vickers Vimy is donated to Brooklands Museum in Surrey.

2009

August	British Airways launches prestige services between London City and JFK with Airbus A318s using old Concorde flight numbers.
23 November	Air France launches first trans-Atlantic Airbus A380 Super Jumbo service, between Paris and New York.

2017

16 June	Norwegian Air International makes first trans-Atlantic flight with Boeing 737-800 between Edinburgh and Stewart International, New York; this is followed on 15 July by the first Boeing 737 MAX 8 flight (named *Sir Freddie Laker*) from Edinburgh to Hartford International, Connecticut.

2018

15 January	A new record for the fastest trans-Atlantic flight in a subsonic aircraft is set by a Boeing 787-9 Dreamliner, G-CKHL, operated by Norwegian. The flight from JFK to Gatwick takes five hours thirteen minutes, three minutes faster than the previous record set by a British Airways flight in 2015. Flight DY7014, which has 284 passengers onboard, is under the command of Captain Harold van Dam. The flight benefits from strong tailwinds over the Atlantic Ocean that reach a maximum of 176 knots (202mph). The tailwinds push the aircraft to a top speed of 776mph during the crossing.

2019

14–15 June	Centenary of Alcock and Brown's first-ever non-stop trans-Atlantic flight.

Chapter One

The Big Prize

To Lord Beaverbrook he was the greatest figure who ever strode down Fleet Street. To the German Kaiser he was so much of a danger that a warship was despatched to shell his house on the Kent coast. To Prime Minister Herbert Asquith he was the media baron whose newspapers helped topple his government.

But to British aviation Alfred Harmsworth, the first Viscount Northcliffe of St Peters in the County of Kent, was the individual who probably did more than any other to champion its early development.

He also provided the inspiration and the motivation behind the attempts to cross the Atlantic Ocean non-stop.

It was said that Northcliffe was obsessed with aviation. He went to the international aviation meeting at Reims, France, in 1909, met Wilbur Wright and appointed Harry Harper as the *Daily Mail's* air correspondent. It was a world-first for the title Harmsworth had founded in 1896 and was now turning into the best-selling newspaper on the planet.

Through the cash prizes awarded by the *Daily Mail*, Northcliffe inspired the early aviators on to ever greater efforts. Among them was the £10,000 offered for the first flight from London to Manchester, the £1,000 for the first cross-channel flight and the £10,000 for the first non-stop crossing of the Atlantic.

Recipients of the *Mail's* largesse included such great names of pioneer aviation as Louis Paulhan, Louis Bleriot, John Moore-Brabazon, Thomas Sopwith, John Alcock, Arthur Whitten Brown and Amy Johnson. Between 1910 and 1930 the *Mail* handed out more than £60,000 in prize money. Some of the losers also received consolation awards, such as the £5,000 that went to Harry Hawker and Kenneth Mackenzie Grieve for the gallant failure of their bid for the Atlantic prize in 1919. Three of the winners of *Daily Mail* prizes, Alliott Verdon Roe (Avro), Sopwith and Hawker, became aircraft manufacturers.

Northcliffe was a visionary whose ideas were often ahead of his time. The *Mail's* announcement of a £10,000 prize for the first trans-Atlantic flight in 1913 was greeted with ridicule. It was, after all, made on 1 April. The general attitude was that if it was not an April Fool gag then it must be a circulation-boosting gimmick. The satirical magazine *Punch* offered a similar prize for the first flight to Mars.

But Northcliffe was not joking. He had been aware that France was still well ahead in the development of aeronautical technology – several of the *Mail's* awards had gone to French aviators in French-built machines – and now he was shocked by the huge sums being spent on military aviation by Germany.

To win the latest *Daily Mail* prize the winner would have to fly between any point in the United States, Canada or Newfoundland and any point in Great Britain or Ireland, and do so within seventy-two continuous hours. The competition was open to pilots of any nationality and to machines of British or foreign construction.

In 1913 £10,000 was a good deal of money, probably the equivalent of around £900,000 today. But at the time, it has to be said, there seemed little immediate danger of Lord Northcliffe having to part with the cash.

That, though, is not to downplay what the journal *Flight* called the *Mail*'s 'munificent money prize to encourage aviation'. The paper had already won praise for its generous encouragement of aviation development, but flying the Atlantic was clearly a challenge of a different order to making the first aerial circuit of Britain or the first aerial crossing of the English Channel.

Several attempts to fly the Atlantic had already been made. The invention of the hot-air balloon in the mid-eighteenth century had at least made the idea a little less fantastic than it had hitherto been. In 1859 an American called John Wise built a huge aerostat that he optimistically called the *Atlantic*. He did not get far. His attempt to cross the ocean ended in a crash-landing at Henderson, New York. Thaddeus Lowe was another to prepare a massive balloon, which he planned to fly from Philadelphia. He called it *City of New York*, but it seems his plans to cross the Atlantic were frustrated by the outbreak of the Civil War.

In 1909, American journalist Walter Wellman attempted to reach the North Pole in an airship named *America*. His attempt failed but the following year Wellman set out for Europe from Atlantic City in a rebuilt dirigible called *America II*. It was 228ft (210.5m) long and 52ft (48m) in diameter, and had a capacity of 350,000 cu ft (9,912 cu m). Its two engines each developed about 80hp, with a smaller 10hp unit supplying air to the balloonists inside the airship's envelope. The main engines were mounted transversely in the keel of the airship and each drove two propellers through bevel gears. The aft engine's propellers swivelled to provide pitch control.

The long keel, which extended for almost the envelope's whole length, enclosed the crew's quarters. Slung underneath was a lifeboat stocked with provisions. A novel feature was the 'equibrilator', which was intended to keep the airship at a more or less constant altitude. It comprised thirty cylindrical tanks strung together on a 330ft (304.6m) steel cable. Actually, it was to prove *America II*'s undoing.

On the morning of Saturday, 15 October 1910 Wellman decided conditions were favourable for his departure. He therefore climbed aboard the airship with his crew, Captain Murray Simon (pilot), Melvin Vaniman (engineer), Jack Irwin (radio operator) and mechanics F.B. Aubrey, and Louis Lond.

Contemporary reports speak of *America II* rising from her moorings at Atlantic City 'in the presence of an enthusiastic crew' and heading out to sea, where it disappeared into the fog. During the day a string of messages was received from the airship that seemed to indicate that the flight was processing as intended. The following morning the airship was reported to be off Nantucket and 'going OK'.

But after that nothing was heard from the craft for two days, which caused 'great anxiety' ashore. Then, on the Tuesday afternoon a wireless message was received from the RMS *Trent*. This reported that the *America II* had suffered engine failure and that, in response to a distress signal from her crew, the *Trent* had stood by to rescue the crew. The airship ditched nearby, the crew, plus cat, scrambled into the lifeboat and were hauled aboard the ship. It later transpired that the equibrilator, of which so much had been expected, had prevented the airship from being steered properly and generated serious vibration as it was dragged through a rising sea.

After it was relieved of the weight of its crew, the airship 'rose to a great height and disappeared'. Wellman had ditched halfway between New York and Bermuda. His bid may have failed but he had covered about 1,000 miles (1,600km), a record for a dirigible.

By 1913 contemporary aeronautical technology still seemed to have some way to go before the idea of a trans-Atlantic a flight by heavier-than-air craft could realistically be contemplated. 'The difficulty with the Atlantic passage,' *Flight* noted, 'is the problem of fuel supply.'

That was not all. 'It is imperative,' the journal observed, 'to recognise the fundamental limitations imposed by the elementary mechanics of the problem. There is no aeroplane that has yet been built for a weight of much less than 15lbs [6.8 kg] per horse power empty, not any aero engine with which we are acquainted that consumes less than half a pint of fuel her horse power per hour. It is, doubtless, possible to improve upon both figures, but the Atlantic is scarcely the proper trial ground for such experimental machines, nor for using either new engines or new fuel.'

The magazine calculated that a suitable aircraft would weigh 'about a ton and a half' or 33lb (15kg) per horse power. This was 'much greater' than any contemporary machine could manage.

Yet there were still a few hopefuls who thought it could be done. French-domiciled Paris Singer of the sewing machine family was reported to be considering an entry, and there was talk of an Italian Bossi machine. In the USA, philanthropist Rodman Wanamaker commissioned a multi-engine flying boat called *America* co-designed by Glenn Curtiss and Briton John Cyril Porte. It was to have been flown by Porte and George Hallett.

The Aeroplane also mentioned a single-engined Handley Page biplane, but thought the most interesting entrant was a British-built Martinsyde. Gustav Hamel, winner of the 1913 Aerial Derby, announced his intention to use it for an attempt on the Atlantic crossing, but he was killed shortly afterwards. The journal reported that a former member of its staff had been sent to Newfoundland to locate and prepare a suitable base. The machine was not completed.

Four years of war would bring major advances in aeronautical technology. Aircraft had become far more capable, so much so that the Atlantic crossing now seemed much more feasible. In November 1918, just days after the Armistice, the Royal Aero Club published the rules for the reconstituted *Daily Mail* Atlantic prize.

The flight could be made in either direction and aircraft could start from land or water, but if seaplanes or flying boats were to be used the competitor would have to fly over the coastline before the start or finish. The elapsed time would be determined to be the moment of crossing the coast or touching land.

Entries started arriving almost immediately. Some were serious; others merely speculative, but the first was received on 15 November 1918. It came from the Whitehead Aircraft Company. *Flight* reported that the aircraft was 'nearing completion', that it had a wingspan of 120ft (40m) and an overall length of 65ft (20m). With its four 400hp (300kN) Liberty engines, the aircraft was expected to be capable of about 115mph (184kph).

The plan was for the big biplane to start from Feltham, Middlesex, and top up its fuel tanks in Galway, Ireland. Captain Arthur Payze was the nominated pilot and he was to be accompanied by a co-pilot, navigator and mechanic. The company had been established by timber magnate and aviation visionary James Whitehead, but its liquidation quickly put an end to the venture.

Flight also reported that 'a 1,600hp Handley Page aircraft has been entered in America as well as a 5,000hp Caproni'. It was also said that a group of financiers was considering an entry on behalf of the Aero Club of America. Later entries included a 375hp (281kN) Fairey IIIa seaplane for Sydney Pickles, with Captain A.G.D. West as navigator, and a 350hp (265kN) Short Shurl for Major J.C.P. Wood.

It was also reported that a Boulton and Paul biplane powered by two 450hp (338kN) Napier Lion engines had been entered. Said to be capable of cruising at 116mph, which would make it the fastest aircraft entered, it was claimed to have a range of 3,850 miles (6,160km). It was later withdrawn, but it is possible that the aircraft existed only on paper. The Alliance Seabird was also Napier-powered and its most obvious novelty appeared to be its enclosed cockpit.

Swedish–American Hugo Sunstedt was reported to be preparing a seaplane for the contest powered by a pair of six-cylinder Liberty engines. A further possibility was said to be a four-engine RAF Felixstowe Porte Fury flying boat, but, according to *The Aeroplane*, the entry had met with 'official opposition'.

More serious were the entries from Handley Page, Sopwith and Martinsyde. The biggest of the three aircraft was the Handley Page V/1500, which was powered by four engines mounted in back-to-back pairs. It was commanded by fifty-five-year-old Rear-Admiral Mark Kerr with, as pilot, Major Herbert Brackley, a former student of John Alcock; Major Trygvie Gran, who had been a member of Robert Falcon Scott's ill-fated Polar expedition, as navigator; and a chief mechanic named Wyatt.

During the First World War the Martinsyde company had become Britain's third largest aircraft manufacturer, with flight sheds at Brooklands and a large factory at nearby Woking. It had been formed in 1908 by H.P. Martin and George Handasyde and was known initially as Martin and Handasyde. By 1914 the enterprise had grown into a successful aircraft manufacturer and a year later it was renamed Martinsyde Ltd.

The Martinsyde Raymor – a contraction of the surnames of its two crew members, Raynham and Morgan – was based on the company's Buzzard A. Mk 1 modified for the trans-Atlantic attempt. Like the Sopwith Atlantic, it was a single-engine, two-seat, single-bay biplane. It was smaller than the Sopwith so, although it was powered by a less powerful Rolls-Royce Falcon III engine developing 285hp (213kW) to the Eagle's 375hp (281kW), it was lighter and could reach 110mph. With a fuel capacity of 393gal (1,480lit), a range of 2,750 miles (4,400km) was claimed.

Work had started on the Sopwith Aviation Company's entry in January 1919 when W.G. Carter began the design of a single-engine, two-bay biplane. It was based on the company's B1 experimental bomber for the Royal Navy. Although only two had been built, one had been used for bombing raids in France. The new aircraft, called the Atlantic, was built in just six weeks and was flying by the end of February.

It bristled with innovations including a wheeled undercarriage that could be jettisoned to cut drag and save weight, built-in wooden skids for landing, an upturned lifeboat that formed the aft decking of the deepened rear fuselage, a wireless powered by a retractable wind-driven generator and staggered seats for the two occupants seated behind the wings.

The fuel tank contained 330gal (1,500lit) of petrol. Power was provided by a 375hp Rolls-Royce Eagle VIII water-cooled V12 engine. Although a four-blade propeller had been used on early tests at Brooklands, a two-bladed 9ft 6in (2.90m) unit was substituted for the actual trans-oceanic flight. The resulting modest reduction in top speed was thought to be worth the gain in thrust at take-off.

To fly the aircraft T.O.M. Sopwith had chosen thirty-year-old Australian Harry G. Hawker as pilot with Lieutenant Commander Kenneth Mackenzie Grieve as navigator. Hawker, dark-haired, lean and with a ready smile, had worked with Sopwith during the war as a test pilot. Back injuries sustained in flying and motor racing accidents had exempted him from military service. The tall, cadaverous Grieve had joined the Royal Navy at the age of fourteen but now, at thirty-nine, he had no previous flying experience.

After testing at Brooklands, the aircraft was dismantled, packed into two big crates and a few small cases and shipped to Newfoundland. It arrived at Placentia Bay on 28 March. Because of ice in the harbour the crates were offloaded from the ship and completed the rest of the journey by rail.

The aircraft was soon assembled, but the weather remained poor. An advance party had already chosen an airfield near St John's and erected a timber shed there, but, as Hawker later recalled, 'the whole place was under snow'. That was far from being the only difficulty. Hawker observed that the area was 'the last place in which one would look for spacious landing grounds'.

If finding a space big enough and smooth enough was one difficulty, an ever-present one was the wind. 'We certainly couldn't afford to risk damaging the aeroplane by attempting trial flights under bad conditions,' Hawker observed. 'There wasn't a day on which we could have made a flight with any degree of certainly about a landing.'

The chosen field was an L-shaped piece of ground about 450ft (415m) above sea level and 6 miles (9.6km) from St John's. The longest stretch, which ran roughly east–west and went past a 200ft (185m) hill, was 400yd (123m) long. The shorter one, 200yd long, ran uphill from the south. There were trees all around but precious little drainage; this meant a soft surface in parts. A gang of sixty men was recruited to fill the worst of the dips.

It had taken about a week to remove the Sopwith Atlantic from its crates and assemble it. The engine had to be regularly run up, which meant that the cooling system had to be drained and refilled each time to prevent the water in the radiator from freezing. Hawker and Grieve were still determined to be first away. That meant the aircraft had to be ready 'in at most a couple of hours' notice', Hawker noted later.

The crew's safety equipment also had to be ready for use. Hawker wrote:

> We had a certain amount of fun in unshipping our boat and testing it on the lake when it behaved very well indeed and in the same water we gave our suits a trial run. There were thoroughly waterproof, coming tight round the wrists and round the neck where they were attached to a strong yoke and being lined with kapok. They were pretty nice and warm and under test they showed their ability to keep out water pretty well. They had originally been filled with kapok but we took the material out and added air bags instead which were lighter and could immediately be blown up by mouth.

As subsequent events were to show, this was a wise precaution. But the two aviators were finding that the enforced idleness was getting on their nerves. Hawker noted: 'It soon became clear that we should have to possess our souls in patience.' And, besides, the competition was catching up.

The Martinsyde entry had reached St John's on 20 April. Its pilot was to be Frederick Raynham, who had gained his aviator's certificate in 1911. He was just seventeen and had been flying a Roe biplane at Brooklands. Later he did much of the test and development work on the Avro 504 and during the First World War worked as a test pilot for Martinsyde. His navigator, Charles Fairfax Morgan, claimed to be a direct descendant of the notorious pirate, Henry Morgan. 'Fax' Morgan had served with the Royal Naval Air Service during the war and had been shot down over France. He spent the rest of the conflict as a prisoner of the Germans. He had lost his left leg, which had been amputated and replaced by an artificial one. But then, as John Alcock would say, none of them intended to walk across the Atlantic.

On their arrival in Newfoundland, Raynham and Morgan soon discovered, as Hawker and Grieve had done, that there was a serious lack of facilities for aircraft operation. They chose a makeshift site near Quidi Vidi Lake, half a mile from St John's. Admiral Kerr and his crew, who had arrived in mid-May, also had difficulty in finding a suitable site. They eventually plumped for an isolated field for their big V/1500 near Harbour Grace, 60 miles (96km) from St John's.

A description of this strip written by one of the waiting reporters showed just how difficult the conditions were:

> It wasn't one field but a series of gardens and farms with rock walls between them. All of these had to be removed, as did three houses and a farm building. A heavy roller, drawn by three horses and weighed down with several hundred pounds of iron bars, eliminated the hummocks. The result, after a month, was a bumpy aerodrome.

As they waited for the weather to improve, the competing crews soon found themselves watching each other 'like a cat watches a mouse'. Hawker and Raynham already knew each other because, before the war, the boyish-looking twenty-six-year-old Raynham had run Sopwith's flying school at Brooklands, while Hawker had been one of his first students.

The aviators kept themselves amused by playing tricks on each other and on the waiting newsmen. On one occasion Hawker and Raynham slipped an evil-smelling codfish into the bed of a reporter whose nocturnal typing had kept everyone awake. On another they poured water into the deep leather chairs of the smoking room at the Cochrane Hotel, which most of the would-be trans-Atlantic aviators had adopted as their headquarters.

Despite prohibition in Newfoundland, there was some solace to be found in alcohol, if not for the teetotal, non-smoking Hawker. 'But it soon seemed hardly good enough,' he wrote later. 'After all, we were on the same errand, we had both got to do our best for British aviation and it was going to be no good being on tenterhooks the whole time – waiting was bad enough for all of us anyhow.'

To ease the tension the Sopwith and Martinsyde crews had agreed to give each other a couple of hours' notice of their departure when they alerted the local wireless stations. That made the atmosphere 'a little less electrical', according to Hawker. It did nothing, however, to improve the weather, which remained, in Hawker's words, 'uniformly rotten'. Whenever the wind blew from the east thick white fog rolled in from the sea and, despite the urge to leave and start the race for the £10,000 prize, the two crews remained on the ground waiting for the weather to improve.

On Friday, 16 May the waiting aviators heard that the three NC flying boats prepared for the US Navy's ambitious plan to fly from New York to England by way of the Azores and Lisbon had arrived at Trepassey, at the south-eastern tip of Newfoundland's Avalon peninsular.

Although they were not entered for the *Daily Mail* prize, the American airmen were still regarded by their British counterparts as competitors in the race to cross the Atlantic by air. National pride was at stake. The stage was set for the most spectacular and most hazardous flight yet undertaken since the invention of the aeroplane just sixteen years earlier.

The Atlantic Contenders

1914

Curtiss-Porte America, two 200hp (150kN) Curtiss OX, completed for R. Wannamaker, flew but did not attempt crossing, co-designed by Lieutenant J.C. Porte RN

Handley Page L.200, one 200hp Salmson, enlarged version of 1914 Anzani biplane, not completed

Martinsyde two-seater, one 215hp (161kN) Sunbeam, undercarriage intended to be dropped after take-off with skids for landing, construction of two examples proposed but not completed

1919

Alliance-Napier Seabird, one 450hp (338kN) Napier Lions, completed after prize had been won, flew non-stop to Madrid with J.A. Peters (designer and pilot) and Captain W.R. Curtiss (navigator)

Boulton Paul P8, two 450hp Napier Lions, did not compete

Fairey IIIa, one 365hp (100kN) Rolls-Royce Eagle VIII, long-range tanks mounted under bottom wing, built but did not start, Sydney Pickles named as pilot

Handley Page V/1500, four 365hp Rolls-Royce Eagle VIIIs, crew Major Brackley, Major T. Gran, Admiral M. Kerr, toured USA after Vimy won prize

Martinsyde-Rolls Raymor, one 265hp (199kN) Rolls-Royce Falcon III, F.P. Raynham (pilot), Captain Morgan (navigator), crashed on take-off in Newfoundland

Short Shurl, one 365hp Rolls-Royce Eagle VIII, intended to fly east–west but came down in Irish Sea on way to Ireland, Major C.P. Woods (pilot), Captain C.C. Wyllie (navigator)

Sopwith Atlantic, one 365hp Rolls-Royce Eagle VIII, flew 1,120 miles in twelve hours but came down in sea with coolant trouble, H.G. Hawker (pilot), K. Mackenzie Grieve (navigator)

Vickers Vimy, two 365hp Rolls-Royce Eagle VIIIs made first non-stop crossing, J. Alcock (pilot), A.W. Brown (navigator)

Not competing for *Daily Mail* prize.

Curtiss NC-4, four 400hp Liberty, first aircraft to cross Atlantic

Chapter Two

Unadulterated Luck

It could be said that honours were just about even in the race to be first to fly the Atlantic Ocean.

It was an American flying boat with an American crew that actually accomplished this feat, but the first non-stop flight was made by a British crew – albeit with one member born to an American family – in a British-built aircraft powered by British engines.

The distinction is important. The US Navy's Curtiss NC-4 with its crew of six took more than three weeks to fly from Long Island, New York, to Plymouth by way of Newfoundland, the Azores and Lisbon, while John Alcock and Arthur Whitten Brown made it to Clifden, Ireland, in a single hop.

A chain of warships stationed across the ocean was intended to ensure the US aviators did not lose their way and could be rescued if they did. This provided a further contrast between the American and British approaches. Harry Hawker, whose bid to make the crossing ended in failure, noted: 'Ours was a private enterprise.'

Indeed it was. Hawker and the other hopefuls were keen to win a prize donated by a newspaper, while the US attempt to fly the Atlantic was treated like a military operation. More than sixty warships, including the battleships *Utah*, *Florida*, *Arkansas*, *Wyoming* and *Texas,* were involved in ensuring the flight was a success. The British Admiralty, however, said it could not spare any Royal Navy vessels to assist the trans-Atlantic aviators.

The NC-4 took fifteen hours eighteen minutes to fly between Newfoundland and the Azores. About a fortnight later John Alcock and Arthur Whitten Brown took sixteen hours twelve minutes for their non-stop 2,270-mile (3,630km) flight between St John's Newfoundland and Clifden, Ireland. Yet, while Alcock and Brown returned home to a heroes' welcome and were immediately knighted by the King, Lieutenant Commander Albert C. Read USN and his crew had to wait ten years for Congress to vote sufficient funds for a special commemorative medal to be struck in their honour. Four presidents came and went before the six flyers were invited to the White House to receive their medals.

As a result, the US public had largely forgotten the achievement of the NC-4 and her crew. It was to be another eight years before a further trans-Atlantic flight by an American citizen really captured the public imagination. In fact, most assumed Charles Lindbergh had been the first aviator to cross the Atlantic.

Perhaps the British were more air-minded than the Americans in the immediate post-war period. The newly created Royal Air Force had ended the war as the world's most powerful air arm and the contribution of aircraft manufacturers such as Sopwith and Vickers to the recent victory was widely recognised.

As a result, the British public, eager to forget the four years of war it had just endured, feted Alcock and Brown as the first heroes of the post-war world in which, it was hoped, conflict would be abolished. Of course, Britain respectfully recognised the feat of Read and his crew. In fact, the town of Plymouth dedicated a plaque to commemorate the event and it took fifty years for this to be emulated by Rockaway in the New York Borough of Queens, which had been chosen as the starting point for the flight.

It has been suggested that the degree of planning and amount of investment involved in the US Navy's Atlantic crossing was on a similar scale to the moon landings half a century later. It is partly for this reason that history has tended to belittle the US achievement in comparison with that of Alcock and Brown. But the Atlantic weather would ensure that success for the US Navy fliers would not come easily.

According to Harry Hawker, the US authorities, 'had made up their minds that to be the first country to send an aeroplane from America to Europe was an honour well worth a big national effort'. But this, he insisted, did not remove any of the gloss from the American achievement. The consequence was that maintaining British aviation prestige was left to privately owned enterprises such as Sopwith, Martinsyde and Vickers.

In fact, though, US seaplane pioneer Glenn Curtiss had his eye on the prize when, in 1914, he designed a flying boat, named *America*, to contest the *Daily Mail* trans-Atlantic competition. The attempt was abandoned with the outbreak of the First World War. But in 1919 the Navy decided to enter the NC (Navy Curtiss) flying boats that had been designed during the war as anti-submarine aircraft.

The idea had originated from within the Navy; Commander John H. Towers, even before the Armistice, proposed that the service should attempt the crossing. His ideas caught the imagination of the Navy Department, where the assistant secretary of the Navy, Franklin Delano Roosevelt, pressed his boss, secretary to the Navy Josephus Daniels, to sanction the flight. Politics intervened. The Navy was battling to retain its own air arm in the face of strenuous efforts to create a separate air force as Britain had recently done. Clearly, making the first trans-Atlantic flight would do the Navy's case no harm at all.

Whatever the motivation, the fact remains that it was a US aircraft and crew that first flew the Atlantic, even if they have not necessarily received full credit for the achievement. Despite the huge fleet of warships stationed below them, the Atlantic weather still ensured that the US Navy air crews were forced to rely on their own skill, courage and resourcefulness.

Lt Richard E. Byrd, later to make the first flight over the North Pole, had also seen the potential of the NC boats and sought permission to be involved with their development. In 1918 he was posted to Canada to command US naval air forces there. After the war, Byrd's expertise in aerial navigation resulted in his appointment to plan the flight path for the Atlantic crossing and to seek suitable locations for the handling of large flying boats.

But the NC boats encountered trouble from an early stage. On 27 March NC-1's right wing had sustained storm damage and it received the corresponding component from NC-2. This aircraft's engine layout had proved unsatisfactory and it was again called upon to supply further parts for NC-1. Two days after the aircraft arrived at Rockaway Naval Air Station, New York, on 3 May its left wing was damaged by fire in a hangar. This series of mishaps cast a pall over the endeavour and there was another when one of NC-4's crewmen accidentally put his hand into the arc of a propeller during an engine test.

This meant that the incomplete NC-2 was not part of the fleet attempting the oceanic crossing. On the morning of Thursday, 8 May, NC-1, NC-3 and NC-4 took off from Jamaica Bay bound for Halifax, Nova Scotia, on the first leg of their epic journey. Appropriately, the formation was led by John Towers, who was also commanding officer and navigator of NC-3. NC-4 was commanded by Lieutenant Commander Albert C. Read and NC-1 by Lieutenant Commander Patrick N.L. Bellinger. Before departure the crew members were presented with a four-leaf clover for luck.

The fleet had reached a point off Cape Cod, the hook-shaped peninsula that curves northwards from Massachusetts, when NC-4's centre engine failed. The aircraft landed on the sea and taxied to the Naval Air Station at Chatham for repairs. NC-3 and NC-1 flew on and arrived at Halifax without further incident. But next morning an inspection revealed serious cracks in their propellers and a day was lost replacing them.

On the 10th, NC-1, and NC-3 continued on to Trepassey, Newfoundland, the planned starting point for the trans-Atlantic flight. The name originates from the French word *trepasses* (dead men) – there is a Baie de Trepassey on the Brittany coast – and a decade later America Earhart would leave

Trepassey on the journey that would make her the first woman to cross the Atlantic by air.

A fleet of warships had been assembled at Trepassey to support the Navy's aviators. A total of twenty-one destroyers were stationed at 50-mile (80km) intervals between Cape Race, Newfoundland and Corvo, the westernmost island of the Azores, to provide visual and radio navigation aids and communication links, as well as weather intelligence and, if necessary, a rescue service.

This might give the impression that the flight was a very simple affair. But in 1919, aerial navigation across a huge trackless ocean was not yet an art, much less a science; aircraft radio was primitive and unreliable, and many of the flight instruments that aviators would later take for granted had yet to be invented. It was certainly not a simple matter to pinpoint nine tiny islands scattered over several hundred square miles of ocean. The fact was that if a pilot flying eastwards missed the Azores the next landfall was Africa, hundreds of miles away.

Meanwhile, at Chatham Naval Air Station, repairs on NC-4 were complete but the flying boat was kept at its mooring by gale-force winds and rain. Accordingly, NC-4's crew were worried that if Towers received a favourable weather forecast he would feel obliged to take advantage of it and head for the Azores without them. Newspapers were now calling NC-4 'the lame duck' and circulating ill-founded rumours that she would be withdrawn from the flight. By the 14th the weather had cleared and NC-4 took off for Halifax, arriving at Trepassey the following day.

Towers had indeed received a favourable weather report on the 15th. He wrote in his log: 'Sun came out, weather is clearing.' He reluctantly decided to go without NC-4, just as its crew had feared. But NC-3 and NC-1 proved to be so overloaded with fuel that they couldn't leave the water. The forecast for the 16th was even better and all three aircraft could go together.

The crews prepared for the long fight in open-cockpit aircraft by pulling leather flying suits over their wool uniforms. Some had chosen to don an additional layer with woollen long johns. Their leather helmets incorporated intercom headsets to enable the aviators to communicate above the noise of the engines.

On the evening of Friday, 16 May the stage was set for one of the most epic journeys in aviation history. The aircraft were now facing the longest and most demanding leg of the trans-Atlantic flight, covering some 1,380 miles (2,208km). An evening departure was necessary if the fleet was to reach the Azores after sunrise next day and enable the aircraft to land in daylight. The three birdcage-like NC boats roared in turn over the water of Trepassey harbour and headed into the gathering darkness over the Atlantic.

The night passed without incident and the aircraft flew over the destroyers on schedule and with reassuring regularity. But during the night, the three aircraft became separated. They had adopted a looser formation to avoid the risk of collision but each aircraft had its own flying characteristics and cruising speed: NC-4 was the fastest of the trio and NC-1 the slowest.

Trouble arrived with the dawn. Sunrise was closely followed by the onset of fog. Through the gathering murk Towers in NC-3 sighted a ship on the horizon that he took to be one of the supporting destroyers. He altered course accordingly. But it was the cruiser *Marblehead* returning from Europe and this mistake was compounded by the erroneous bearing, which was to take NC-3 far off course.

With fuel running low, Towers' dead reckoning convinced him that he was somewhere near the Azores. He decided to alight on the sea to allow time for a navigation sight to be taken. But a high sea was running and the landing was so rough that the impact broke the struts supporting the aircraft's centre-line engines. NC-3 was so badly damaged that it could not take off.

Bellinger, in NC-1, was having similar difficulty with navigation, but he managed to land the aircraft intact. However, once down it could not take off through the 12ft (11m) waves now running. Indeed, the craft would be lucky to survive them.

Read had become virtually lost in the fog. At one time it was so thick that the crew could not see from one end of their aircraft to the other. The pilot became so disoriented that he almost put the big machine into a spin. But then, Ensign Herbert Rodd, the radio officer, was able to pick up radio bearings and weather information from the supporting destroyers hidden below in the fog.

After more than fifteen hours in the air, a combination of Read's dead reckoning and Rodd's radio reports convinced the crew that NC-4 was close to its destination. All hands were ordered to keep a sharp look-out. Then, through a small break in the fog, the green of an island was sighted. It was Flores, one of the western Azores.

Using it as a checkpoint, Read swung NC-4 eastward for the islands of Fayal and Sao Miguel. The fog thinned but soon thickened again and Read settled for the immediate haven offered by Fayal. NC-4 landed in the harbour of Horta a little before noon. It was none too soon: within minutes a great bank of fog blotted out the port completely.

When he boarded the cruiser *Columbia*, base ship for the NCs at Horta, the first thoughts of Read and his crew were for their comrades aboard NC-3 and NC-1. It soon became apparent that NC-1, trapped and pummelled by great waves, was lucky to stay afloat let alone take off. But then the Greek freighter *Ionia* loomed out of the fog and rescued Bellinger and his crew. Attempts to salvage the derelict NC-1 were thwarted by the heavy seas and it sank three days later.

Of NC-3 there was no word. In fact, its fate would remain a mystery for forty-eight hours. Before leaving Trepassey, Towers had jettisoned the emergency radio transmitter to reduce weight for take-off. This meant that NC-3 could receive radio calls but not make them. Its inability to take off transformed the craft from flying boat to just boat. Only pure seamanship could save NC-3 and its crew. Towers calculated that within two or three days NC-3 would drift in close to the island of Sao Miguel in the eastern Azores. His estimates were proved correct when, on 19 May, NC-3, battered and almost derelict, reached the harbour of Ponta Delgada after an epic 250-mile (400km) journey.

NC-4 was now alone and the success of the enterprise rested on the shoulders of Read and his crew. Their further progress, however, was frustrated by the weather. The craft rode its moorings at Horta, kept there by high seas, rain and fog. On the 20th, the weather cleared sufficiently to permit a take-off. In less than two hours NC-4 reached Ponta Delgada. Read planned to leave for Lisbon the next day, but there was more delay. Weather and engine troubles delayed the departure for a week.

On Tuesday, 27 May NC-4 was ready to take off for Lisbon and the European mainland. Lieutenant James L. Breese and Chief Machinist's Mate Eugene S. Rhoads checked the engines, while Herbert Rodd ensured his radio set was ready. On Read's command, Lieutenant Elmer Stone advanced the throttles to send the big flying boat roaring across the harbour, trailing a huge V-shaped wake of spray. The aircraft lifted off at 08:18 hours.

There was another chain of fourteen destroyers between the Azores and Lisbon and as NC-4 passed overhead each radioed the news to the base ship *Melville* at Ponta Delgada and the cruiser *Rochester* in Lisbon. These ships relayed the news to the Navy Department in Washington. Finally, word came from the destroyer *McDougal*, the last ship in the picket line, that completion of the flight was only minutes away.

In NC-4 all eyes were focussed eastwards where the horizon was fading into the deep purple of twilight. Then at 19:39 hours, from the centre of the darkening line, a spark of light could be discerned. It was Cabo da Roca lighthouse. Europe's most westerly point had been sighted. Minutes later, NC-4 roared over the rocky coastline and turned south for the Tagus estuary and Lisbon. The actual flying time from Ponta Delgada had been nine hours forty-three minutes and from New York twenty-six hours forty minutes.

Read, a quiet New Englander, who at just 5ft 4in (1.6m) in height seemed the most unlikely looking hero of the group, was said to be a man of few words. But he would describe this moment as 'perhaps the biggest thrill of the whole trip'. Each man on board, he reported later, realised that, 'no matter what happened—even if we crashed on landing—the trans-Atlantic flight, the first one in the history of the world, was an accomplished fact'.

Indeed it was. At 20:01 hours on 27 May NC-4's keel sliced into the waters of the Tagus. It was one of the key moments in aviation history. Read later wrote to his wife that the successful flight represented 'a continuous run of unadulterated luck'.

All three NC crews were reunited in Lisbon, where they received a rapturous welcome from the city and the Portuguese government. But the real objective of the journey still lay ahead. On the morning of 29 May, after a two-day stay in Lisbon, NC-4 left for Plymouth. Another picket line of ten ships had been stationed along the route, but a few hours later NC-4 was down. Engine trouble forced the aircraft to land on the sea off the Monedego River. The engine was soon repaired but Read was not prepared to risk landing at Plymouth in darkness so the night was spent at El Ferrol, Spain.

The next day NC-4 completed its journey. The flying boat landed in Plymouth harbour early in the afternoon of 31 May, having received an escort for the final stages by three Royal Air Force Felixstowe F.2A flying boats, which were dwarfed by the American aircraft. The leading RAF machine flew the Stars and Stripes between its port interplane struts with the Union flag fluttering from the opposite site.

The NC-4 made a celebratory fly-by before landing at 14:27 hours to an enthusiastic welcome. When it landed, Read and his crew were taken by launch to the cruiser USS *Rochester*, where the band was playing *America Forever*. They were then conveyed to the depot ship USS *Aroostook*, where they were able to change out of their flying clothes and into clean uniforms.

Later, they went ashore at Plymouth, climbing the steps of the monument built to commemorate the Pilgrim Fathers' departure 299 years earlier. They were welcomed by the city's mayor and there was a reception in their honour at the Grand Hotel. The next day Read and his crew travelled by train to London, where they were greeted by scores of American soldiers and sailors, who insisted on pushing their car all the way from Paddington Station to the Royal Aero Club and another reception.

During their stay in London, Read met Harry Hawker, newly rescued from the sea after his failed attempt to cross the ocean. Their next stop was Paris, where the peace talks were in progress at Versailles. President Woodrow Wilson told them: 'The entire American nation is proud of your achievement.' The flyers also met British Prime Minister David Lloyd George.

Back in London they met the Prince of Wales and Winston Churchill as well as senior representatives of the British aircraft industry. At the Air Ministry Read and Towers received the Air Force Cross, while Eugene 'Smoky' Rhoads was awarded the Air Force Medal to recognise his crucial role in the success of the flight.

On 15 June the flyers boarded the USS *Zeppelin* at Brest for their return journey to New York. They were welcomed by a delegation of naval officers and others, but there was no official welcome. There was no ticker tape parade along Broadway, either. Perhaps the laconic Read had not helped. In New York he had disappointed reporters by saying: 'My flight for the most part was quite a monotonous affair. It was simply doing the same things over and over again from start to finish.'

Flight conceded that Read and his crew deserved congratulations for their achievement. They would go down in history as 'the first men to cross the Atlantic in the air'. Yet the flight had been 'rather a triumph of organisation than anything else,' adding, 'it is to the American naval authorities no less than to Commander Read and the crew of NC-4 that the success of the cross-Atlantic flight is due.'

The Times mused on the differences in approach of Britain and the USA. The paper thought that the official British attitude towards such undertakings was based on a tradition 'of leaving pioneer work to private endeavour with a substantial reserve of disapproval in case of failure'. This, the paper noted the day after Alcock and Brown had completed their flight, went back to the days of 'that gaunt patroness of official meanness, Queen Elizabeth'. It added: 'The Americans with NC-4 have shown us that there is another way and a very efficient way.'

On 30 June the NC crews were met by Navy Secretary Josephus and Assistant Secretary Roosevelt for handshakes and photographs. Josephus told the flyers that Congress planned to mark their achievement with a special medal struck in their honour. It was to be eleven years before President Hoover pinned the decorations on the chests of the small continent of trans-Atlantic flyers who visited the White House in May 1930.

Three of the aviators, including John Towers the instigator of the flight, became admirals. Marc Mitscher, originally appointed to pilot the ill-fated NC-2 but moved to NC-1, commanded USS *Hornet*, from which, in 1942, an audacious bombing raid on Tokyo was launched. Later he commanded the Fast Carrier Task Force. In the early 1920s, Mitscher had played an important role in assisting the Navy retain its own air assets by appearing for the prosecution at the court martial of General Billy Mitchell, who had taken his case for an independent air force to the public via the press.

On its return to the US, NC-4 completed a celebratory tour of the eastern and southern seaboards by flying up the Mississippi to St Louis, where it was handed over to the Smithsonian Institution. Later it was donated to the Naval Air Museum in Pensacola, Florida, where it remains on display.

The first trans-Atlantic flight had occupied twenty-four days. During this time the world had held its breath and the story occupied the front pages of

American newspapers. But by the time the flyers returned home the story had moved on. Alcock and Arthur Whitten Brown had completed their non-stop flight and become the new heroes. But for all that they had not been the first to show that the ocean could be crossed by air. That honour belongs to Lieutenant Commander Albert C. Read, his crew of five, and the United States Navy's NC-4.

Less obvious were the other achievements. The US aviators' flight had marked the first use in aviation of the radio compass, air-to-ground, air-to-air and intercom radio systems, the bubble sextant, wind and drift indicator and the Great Circle air route to Europe.

In 1957 James Breeze, NC-4's engineering officer, was flying to Europe when the flight attendant asked him if it was his first trip across the ocean. 'No,' he replied, 'I flew across once before – thirty-five years ago.'

The First Trans-Atlantic Aviators

Seaplane Squadron 1 was commanded by Commander John Towers. The crews of the aircraft were:

NC-1

Lieutenant Commander Patrick N.L. Bellinger, CO, pilot and navigator; Lieutenant Louis T. Barin, co-pilot; Lieutenant (junior grade) Harry Sadenwater, radio officer; Chief Machinist's Mate Rasmus Christensen, engineer; Chief Machinist's Mate C.I. Kesler, engineer

NC-2

Lieutenant Commander Marc A. Mitscher, CO (transferred to NC-1)

NC-3

Commander John H. Towers, CO; Commander Holden C. Richardson, pilot; Lieutenant Commander Robert A. Lavender, radio officer; Lieutenant David H. McCulloch, co-pilot; Chief Bosun's Mate Lloyd R. Moore, engineer

NC-4

Lieutenant Commander Albert C. Read, CO and navigator; Third Lieutenant Elmer F. Stone (US Coast Guard), pilot; Lieutenant (junior grade) Walter Hinton, co-pilot; Lieutenant James L. Breese, co-pilot and engineer; Ensign Herbert C. Rodd, radio officer; Chief Machinist's Mate Eugene S. Rhodes, chief engineer.

The Nancy Boats

It was appropriate that the NC-4 flying boat should have been designed and built by Curtiss as Glenn Curtiss was a pioneer of water-borne aircraft and his aircraft, named *America*, had been built in 1914 to contest the *Daily Mail* trans-Atlantic competition.

The NC-4 was one of four NC (Navy-Curtiss) flying boats, designed originally to help protect American shipping against attacks by German U-boats, although they were too late to see service in the First World War. Accordingly, the aircraft were designed from the outset to have trans-Atlantic range.

The Navy's requirement was drawn up in September 1917 and design studies were conducted by a team of Navy engineers. Curtiss was appointed as a consultant to work out the design's details. In January 1918 Curtiss was contracted to produce four flying boats. They featured a 45ft (13.72m) hull of advanced hydrodynamic design and were built by the Herreshoff Manufacturing Company of Bristol, Rhode Island.

The hull was constructed of double-planked spruce on a mahogany frame with muslin and shellac between the two layers. The biplane tail with triple fins and rudders was carried on three booms projecting from the top wing and the rear of the hull. It was braced by a transverse triangular structure. The first aircraft was powered by three 400hp (300kN) Liberty engines, mounted between the wings.

Large aircraft for their day, just 1ft (0.92m) shorter than a Second World War Lancaster bomber, the four boats were numbered separately, NC-1 to NC-4. But the war was ending as flight testing of NC-1 began. It flew for the first time on 4 October 1918 and on 25 November gave a convincing demonstration of its load-carrying ability by taking fifty-one people on a single flight from Rockaway Naval Air Station to set a world record.

Yet the aircraft was considered under-powered for the trans-Atlantic attempt and construction of the three later aircraft was delayed by the addition of a fourth engine. NC-2 had its power plants mounted in tandem pairs but this was found to be an unsatisfactory configuration and in NC-3 and NC-4 the fourth engine was mounted behind the middle unit driving a pusher propeller.

After the Armistice the Naval Aircraft Factory at Philadelphia completed six more NC-type flying boats. Built initially as tri-motors, four were later converted to an NC-4-type four-engine configuration, although the other two aircraft had been lost. The converted aircraft served during 1920–22 with the US Navy's East Coast Squadron before being retired.

Curtiss NC-4 Specification

Crew	six
Length	overall 68ft 5.5in (20.8m); hull 44ft 9in (13.8m)
Wingspan	126ft (38.4m)
Height	24ft 5in (7.4m)
Loaded weight	28,000lb (12,700kg)
Engine	four Liberty L-12 water-cooled V12s each developing 400hp (300kW)
Speed	maximum 91mph (146kph); cruise 77mph (123kph)
Service ceiling	4,500ft (1,370m)
Range	1,470 miles (2,366km); endurance, 14.8 hours

Chapter Three

A Bit Ticklish

The Avalon peninsula is joined to the island of Newfoundland by a narrow strip of land. On the map it looks like a smaller replica of the main part of the island.

Yet Avalon contains just over half of Newfoundland's population, mainly due to the presence in the north-east corner of St John's, the island's capital. It is one of North America's oldest cities with one of its least inviting climates: St John's is Canada's foggiest, wettest and windiest city. It snows a lot, too.

Newfoundland has played a major role in the development of communications between the old world and the new. The Genoese John Cabot, commissioned by England's King Henry VII, is credited with the discovery of Newfoundland in 1497, while in 1901 Guglielmo Marconi received the first trans-Atlantic wireless signal in St John's.

With its inhospitable climate, lack of suitable fields and absence of facilities, Newfoundland would seem to be the last place a pioneer aviator would choose for a trans-oceanic flight. But it offered one benefit no other locations could match: it is the nearest part of North America to Europe.

That was why in the early summer of 1919 Newfoundland was at the centre of one of the year's biggest news stories. Teams of intrepid aviators representing some of the world's biggest and most powerful aircraft manufacturers were assembling there to attempt something nobody had achieved before.

The challenge, first issued in 1913 and repeated six months later, was the *Daily Mail*'s £10,000 for the first aviator to cross the Atlantic Ocean. The winner would have to fly between any point in the United States, Canada or Newfoundland to any point in Great Britain or Ireland in seventy-two continuous hours.

Although the American flying boat crews were not eligible for the £10,000 *Daily Mail* prize and their planned route across the ocean was a southerly one, their arrival in Newfoundland ratcheted up the tension among the waiting aviators. When they were reported to have reached the Azores the news helped Hawker and Grieve make up their minds. They would get as much sleep as they could because they next night they expected to be going without; they would have something more 'interesting' to do.

At around noon on 18 May, Hawker told Raynham he had decided to go. He and Grieve struggled into their watertight suits, shook hands with onlookers and climbed into their cockpits. They took off at 15:42 hours local time (17:42 GMT) on 18 May. 'Tell Raynham I'll greet him at Brooklands,' Hawker shouted to onlookers. He gunned the engine and the Sopwith, its tanks brimming with fuel, lurched over the bumpy ground.

'Getting off was just a bit ticklish,' Hawker would recall. What he meant was that with the wind blowing about 20mph (32kph) east-north-east he had little choice but to take off diagonally. This meant skirting rising ground and avoiding a drainage ditch running across the end of the field. The going was rough and the Sopwith hopped over the bumpy ground, missing the ditch by inches, but it cleared the trees lining the field by 'a respectable distance'.

As soon as the aircraft crossed the coast, Hawker triggered the undercarriage release and the wheels fell away to the sea. This cut both weight and drag: Hawker noticed that the aircraft's speed had increased by 7mph (11kph). The sky was bright and clear at the start of the journey, but fog was encountered after just ten minutes as the aircraft was still climbing. It did not prevent Grieve from taking a drift reading from the sea.

But soon the airmen found themselves flying in clear weather and this persisted for some hours. Cruising at 10,000ft (3,050m) with everything apparently set fair, the aircraft was pretty well flying itself at 105mph (168kph) with Hawker guiding it along the course Grieve had given him. By 22:00 hours the aircraft was flying above cloud. The sky above them had turned purple. But within the next fifteen minutes the weather had changed 'for the worse'.

The sky became hazy and the downward view was all but obscured by cloud. Ahead the cloud was even thicker but the Sopwith Atlantic plunged into it. The ride became bumpy and there was intermittent rain. But the aviators were warm and dry and optimistic that it was a just bad patch through which they would soon pass.

But at 23:00 hours came the first sign that all was not well with the aircraft's engine. The water temperature was beginning to rise and did not fall as it should have done when Hawker opened the radiator shutters. The clouds appeared to be getting thicker and heavier and the sea was now invisible. The wind direction seemed to have altered, making it more difficult for Hawker to hold his course.

As the cloud appeared 'more formidable' and the going got even bumpier it was clear that the aviators could not climb above the murk without using more fuel than they could afford. Besides, climbing might make the water temperature rise. 'So we just had to go round the clouds as best we could,'

Hawker would observe later, 'but there were so many of them that Grieve never had a chance to take a sight on the stars.'

As he recounted later, Grieve encountered many difficulties in navigating the aircraft. The 'pretty cramped conditions' did not make the taking of star sights 'any easier than it might have been'. The intense cold made removing his gloves hazardous. He wrote later:

> So far as warmth was concerned I had nothing to complain of for our suits kept us quite comfortable until I had to make an observation with the sextant and then conditions were not quite so good as the working of this instrument and the subsequent calculations meant taking one's gloves off. My silk gloves, which I wore next to the skin, managed to get themselves blown overboard ... Whilst messing about with my instruments I suppose I was lucky not to get badly frost-bitten – as a matter of fact I did get a touch or two. What was even more distressing was the fact that my handkerchief got frozen as stiff as a board!

Conditions improved a little later. The moon was now visible. But the engine water temperature was still rising even though the radiator shutters were wide open. Hawker wondered if the filter was being clogged up by a mixture of rust flakes and bits of solder. He decided to shut the engine down and put the aircraft into a dive to try and clear the filter. The aircraft dropped from 12,000ft to 9,000ft (3,660m to 2,740m) and Hawker was gratified to note that at the end of this process the temperature remained 'moderate', even though the aircraft was climbing to regain its previous altitude.

There was little reason for relief, though, because repeating this process would use more fuel. In any case, the aircraft was now heading into the wind, raising fuel consumption still further. At 00:30 hours the aircraft had travelled 800 miles (1,280km) and Hawker was still trying to dodge the worst of the clouds. But every time he climbed the water temperature rose. It was getting perilously close to boiling point.

The cooling system's capacity was 19gal (86lit) but Hawker was well aware that once the water started boiling it would not be long before it evaporated despite the freezing cold. Hawker managed to coax the aircraft back up to 12,000ft. He throttled the engine right back, but he could see that the top wing was covered with ice. Steam from the radiator was spouting out 'like a little geyser from a tiny hole in the middle'.

On the credit side, Grieve was able to continue his observations of the stars and keep the aircraft on course. But by 06:00 hours the aviators were confronted by a bank of very solid-looking black clouds rising to at least 15,000ft (4,570m). Any attempt to gain altitude and fly above the murk caused the water temperature to rise dangerously.

Hawker decided to descend to 6,000ft (1,800m) where, if anything the cloud was thicker, but the water was now boiling. It was then that the aviators had what Hawker would later call 'a very narrow squeak indeed'. When he tried to open the throttle, there was no response from the engine. Hawker shouted at Grieve to start working the petrol pump.

Nothing happened 'except that the Atlantic rose to meet us at rather an alarming rate'. The situation was ironic. 'The whole damned Atlantic was beneath us,' Hawker observed. 'We could have let down a bucket and a rope for water if we had had a bucket and a rope.'

But the situation was desperate. Hawker recalled: 'We were gliding downwind at a pretty good speed, the sea was very rough, and when we hit it I knew very well that there was going to be a crash of sorts, and that if he remained where he was Grieve would probably get badly damaged, as he would be shot forward head first on to the petrol tank.' He added: 'So I clumped him hard on the back and yelled to him that I was going to "land".'

At that moment it must certainly have seemed that this was about to happen. But thanks to Grieve's energetic pumping the engine fired at last. Hawker opened the throttle and hauled back on the control column to put the aircraft into a climb.

The aviators were well aware they had been lucky and had cheated death. But for how much longer? Indeed, it had become clear that the Sopwith Atlantic was now doomed even though it might limp on for another hour or two. The irony of the situation was that, although there was enough fuel to reach Ireland, the engine's cooling water would soon run out.

Accordingly, Hawker and Grieve decided to abandon their attempt and search instead for a ship to rescue them. The weather was worsening with a stronger wind whipping up a rough sea. It was an anxious time, but then a hull loomed out of the murk. 'I'm ready to admit,' Hawker wrote later, 'that I shouted with joy.'

Below the crippled aircraft was the Danish steamship *Mary*. Hawker flew round it and fired three Very light distress signals. The crew appeared on deck, no doubt surprised to see an aircraft this far from land. The Atlantic flew on a few miles past the ship. Then, according to Hawker, he:

> … judged the wind from the wave crests, came round into it and made a cushy landing in spite of the high sea that was running. The machine alighted quite nicely and, thanks to her partly empty tanks, rose clear of the water, although now and then the waves sloshed right over it and us and soon played havoc with the main planes.

The two aviators managed to launch their on-board lifeboat without difficulty and they scrambled into it. The life-saving suits also proved their worth.

Meanwhile, the *Mary's* crew had launched a lifeboat and, despite the distance coming down to just 200yd (180m), it took ninety minutes for Hawker and Grieve to be hauled to safety, such was the state of the sea. They were plucked from the ocean at 20:30 hours GMT, fourteen hours thirty minutes after taking off from Newfoundland.

The *Mary*'s master, Captain Duhl, later recalled how the two airmen were rescued:

> When I came on the bridge, the machine had already alighted in the water. The airmen told us that they had dropped rockets but we did not see them. The work of saving them was pretty difficult because it was blowing very hard. Another hour and we might not have been able to launch the boat. Hawker and Mackenzie Grieve were in water up to their waists but their water-tight suits kept them dry. It was rather a difficult hour before we succeeded in reaching them. All the airmen wanted to do was sleep. When they had had their sleep out and got a good meal with a glass of schnapps they were all right.

Had they not been rescued by the *Mary* the two aviators would probably not have survived for much longer. As it was, they had to leave the mail bag they were carrying, although it did eventually reach Falmouth.

As it happened, the *Mary* carried no radio so the aviators' rescue remained unknown. Back in Newfoundland, concern was mounting: the aircraft had fuel for twenty-two hours' flying, but by the following afternoon nothing had been heard, despite the aircraft carrying a radio. It was presumed that it had come down in the sea and that Harry Hawker and Mac Grieve were lost.

Ever since the trans-Atlantic contest had started, the Air Ministry in London had ordered a round-the-clock watch to be maintained. When Hawker and Grieve left Newfoundland, radio stations and ships in the vicinity were asked to keep a listening watch. When it was clear the aviators had come down in the sea, two RAF squadrons were ordered up to search coastal areas. The Admiralty issued a warning to all ships to look out for any sign of the aircraft.

Even so, it was clear that in future airmen making attempts on the Atlantic crossing could not rely on the support of the Royal Navy. In May 1919 *Flight* reported: 'The government, in view of comments as to responsibilities in this matter, feel bound to call attention to the many and heavy responsibilities of the British fleet.'

Hawker's wife, Muriel, whom he had married in 1917, received a long telegram of condolence from Buckingham Palace. It said that King George V, 'feels that the nation has lost one of its most able and daring pilots who sacrificed his life for the fame and honour of British flying'. But His Majesty's sympathy was a little premature. In a bizarre twist to the story, a memorial service was actually taking place at Hook when a telegram boy arrived with the news that Hawker and Grieve were alive.

Captain Duhl of the *Mary* had hoped to fall in with a ship equipped with a radio, but the intensity of the storm worsened and he was forced to heave to and ride it out. The ship had drifted north and away from the busier shipping lanes, and it was not until the *Mary* was off the Butt of Lewis that the aviators' rescue could be announced.

The ship's siren sounded and flags were used to signal the message 'Saved hands Sop aeroplane'. A Navy communications station replied with the question: 'Is it Hawker?' The aviators were taken on board the destroyer HMS *Woolston* outside Loch Erriboll and were landed at Scapa Flow, where they were transferred to HMS *Revenge,* flagship of the Grand Fleet. There to welcome them was the commander-in-chief, Vice Admiral Sir Sydney Freemantle. The following day, they landed and headed for home. What they could not have expected was the enthusiastic welcome that awaited them. It was almost as if they had succeeded in their attempt, as Hawker observed later.

'As to the reception we received and the demonstrations that were made, need I say that Grieve and I were more than deeply touched,' Hawker recalled. 'We are completely agreed that the whole of this business was utterly undeserved and out of all proportion to what we had tried – and failed to do.'

After a thanksgiving service a few days later, Muriel Hawker's brother rushed by car to London to meet the aviators, apparently being let off a charge of speeding up Putney Hill. Hawker and Grieve were travelling from by train from Scotland where they had landed. They were feted as heroes, waving crowds assembling at intermediate stations with an even bigger throng waiting at King's Cross station. And there was another telegram from the King.

When the heroes attended a celebratory reception, the crowds were so enthusiastic that Hawker had to abandon the Rolls-Royce tourer conveying him to the Royal Aero Club for a police horse. Next day he went to Buckingham Palace, where the King awarded him, a civilian, the Air Force Cross, after which the *Daily Mail* entertained Hawker and Grieve at the Savoy. They also received the newspaper's cheque for £5,000.

Hawker revealed that the King had asked him about 'the public outcry for Admiralty help'. The Australian responded that it had all been 'nonsense so far as we were concerned'. He and Grieve 'had never wanted any such help and had not in the least relied upon receiving it'.

Meanwhile, the Sopwith Atlantic was found afloat in the sea and salvaged by the crew of the US steamer *Lake Charlotteville*. It seemed that the cause of the ditching could now be established. Until then it had been assumed that the engine had overheated because the radiator shutter controls had been reversed.

But that would have meant the overheating would have happened soon after the full-load, full-throttle take-off, not halfway across the ocean. In any case, Hawker was adamant that he had closed the shutters before ditching to prevent sea water being turned into steam and scalding Grieve in the front cockpit.

Hawker had maintained that the source of the trouble had been a choked water pump filter, but examination failed to confirm this, probably, Hawker suggested, because the immersion in seawater had cleared it. But it seemed more likely that ice had blocked the radiator and water pipes. The cutting-out and recovery of the engine was thought to point to carburettor icing, about which little was known in 1919. Alcock and Brown would experience it on their successful crossing.

But why had Hawker and Grieve not been able to use their wireless equipment? Grieve wrote later: 'It was valueless. In fact, we were not able to make any use of it at all.' He explained that the directional equipment had been removed from the aircraft before the flight. The Type 55A set supplied by the Air Ministry was found unsatisfactory in tests in Newfoundland and replaced with a Type 52A. But there had been insufficient time to test it before the Atlantic flight. There had been some interference from the engine's magneto but, worse still, a large four-bladed propeller, which had been intended to generate power for the set, failed to perform as intended.

Grieve also defended the decision not to make a test flight with the replacement equipment before leaving Newfoundland. He said that the aerodrome there was so bad that:

> to make any test flights at all was to take a serious risk with the aeroplane and so far as the flight was concerned we *were* counting on the machine and we were not counting upon the wireless, so that really we took the lesser of two evils.

The damaged Atlantic aircraft went on display on the roof of the Selfridges department store in London's Oxford Street.

Hawker and Grieve's failure had made Raynham and Morgan favourites to take the prize. But Raynham's trans-Atlantic journey was to be even shorter. Departing two hours after Hawker, the heavily laden Martinsyde, brim full of fuel, bounced over the bumpy field slowly at first. As the speed built up after 300yd (275m), a big bump tossed it into the air. The biplane flew for just 100ft (30m), drifting sideways in the crosswind. But it was not flying fast enough and it crashed. The undercarriage was torn off and the nose buried itself in the earth as the horrified onlookers ran across the field to help. Raynham crawled from the wreckage unaided, but Morgan had to be lifted from his cockpit. Both the airmen were injured, but Morgan was sent back to England on medical advice suffering from serious eye injuries.

His replacement as Raynham's navigator, Lieutenant C.H. Biddlecombe, arrived in Newfoundland on 14 June. But by the time the aircraft had been repaired, Alcock and Brown had already made the first successful non-stop trans-Atlantic flight. All the Martinsyde team could now hope to salvage from their efforts was to better the Vimy's time for the crossing of sixteen hours twelve minutes.

Accordingly, Raynham tried again, taking off at 15:15 hours local time on 17 July. This time the aircraft did get off the ground but flew for only about 60yd (50m) before plunging to earth. This time it was completely wrecked but the crew was unharmed. The Martinsyde effort was now at an end. Two attempts, in which it had covered little more than a couple of hundred yards, had ended in abject failure.

Raynham returned to England with a bag of mail from Newfoundland, which was finally posted on 7 January 1920. Two years later, the Martinsyde company went into liquidation following a factory fire. By then it had switched to the manufacture of motor cycles.

But although Raynham and his crew did not return home to a heroes' welcome, Harry Hawker went out of his way to praise them. He wrote:

> The men who should have had the reception are Raynham and Morgan for what they did was a magnificent act of pluck. The east-north-east wind was not by any means a bad one for our getting off, for it suited our aerodrome pretty much as well as any other and better than most but it was almost the worst possible wind for the Martinsyde aerodrome. But knowing this, Raynham and Morgan never hesitated to attempt the flight and in doing so displayed a spirit and courage for which Grieve and I have nothing but the most intense admiration and respect. They were visited with very cruel hard luck indeed.

Once the *Daily Mail* prize had been claimed there seemed little point in any of the other would-be contestants risking their lives in attempting the crossing. The reporters covering the story had moved on from Newfoundland. Admiral Kerr and the crew of the big Handley Page bomber decided to do the same; their attempt on the *Daily Mail* prize had been delayed by the need for replacement radiators. They headed south for New York, but engine trouble forced them to land at Parrsboro, Nova Scotia, on 4 July. The rough landing damaged the aircraft so badly that it took nearly four months and a new fuselage shipped over from England to get it airworthy again. The flight continued to New York on 9 October.

Of the other attempts little remains to be said. A Swedish-American pilot named Hugo Sunstedt had entered a biplane of his own design. Called *Sunrise*, it flew well enough and Sunstedt was an experienced pilot. But during a test flight in February 1919, another pilot spun the *Sunrise* into the sea off Bayonne, New Jersey.

On 8 April, Major J.C.P. Wood became the second competitor to drop out. He took off from Eastchurch, Kent, in a Short seaplane and headed west. Unlike the other competitors, Wood had elected to make the crossing from east to west and flying into the prevailing wind rather than with it. It was probably fortunate, then, that the Short's engine quit 20 miles (32km) from Holyhead, Wales. The aircraft ditched in the Irish Sea, from which Wood and his navigator were duly rescued.

If Wood's attempt had seemed foolhardy the same could be said of other contestants, even Raynham and Morgan. They had decided to go despite adverse weather, but they had been motivated by the departure of Hawker and Grieve. Yet, even in the twenty-first century, making a long oceanic crossing in a single-engine aircraft is best attempted by an experienced crew in a well-equipped aircraft. But then in 1919 none had that kind of experience.

After the premature end of the Sopwith Atlantic's flight, Harry Hawker was adamant that he would be perfectly happy to make another attempt in a single-engine aircraft. After his and Grieve's survival had been confirmed there was much discussion about the 'unnecessary' risks they had run in attempting to fly over 1,800 miles of ocean on just one engine. 'As a matter of fact,' Hawker wrote, 'this is very nearly a complete fallacy and I myself am quite satisfied that so far as reliability is concerned a single-engine aeroplane is a match for any multiple-engine machine.'

Atlantic Men

The Atlantic was the product of one of Britain's best-known aircraft manufacturers. The Sopwith Aviation Company had made a major contribution to the nation's war effort by producing more than 18,000 machines and the 5,700 Camel fighters included in this total helped make the post-Armistice Royal Air Force the world's most powerful air arm.

Thomas Octave Murdoch Sopwith learned to fly in 1910 and received the Royal Aero Club's aviation certificate number 31. That year Sopwith won a £4,000 prize for a flight from England to mainland Europe, during which he flew 169 miles (272km) in three hours forty minutes. He used the prize money to establish a flying school at Brooklands in Surrey.

In 1912 Sopwith and his associates started building aircraft and later that year Australian-born Harry Hawker used the company's first aircraft, a modified Wright Model B, to win an endurance prize, remaining airborne for eight hours twenty-three minutes. The order for the first military aircraft prompted Sopwith to move from Brooklands to larger premises at Kingston-upon-Thames.

In fact, Sopwith established three factories in Kingston. He transformed a small town on the southern outskirts of London into the source of most of Britain's wartime fighters and the heart of the military aircraft industry for years afterwards. Over nine decades Sopwith and its successor firms are thought to have employed 40,000 people in Kingston.

During the First World War Sopwith turned out a variety of successful designs including the Pup, One-and-a-Half Strutter, Dolphin and Snipe. But despite, or perhaps because of, its wartime success, the company was faced with punitive taxes. The decision was taken in 1920 to liquidate it and start a new company.

In this enterprise Sopwith was joined by Hawker, Fred Sigrist and Bill Eyre, each of whom contributed £5,000. To avoid any possibility of claims against the new company in respect of wartime contracts undertaken by the previous one, they named it H.G. Hawker Engineering. As Sopwith put it this would:

> ... avoid any muddle for if we'd gone on building aeroplanes and called them Sopwiths – there was bound to be a muddle somewhere – we called the company the Hawker Company. I didn't mind. He was largely responsible for our growth during the war.

Later the company would become Hawker Aircraft Limited. In 1935 it merged with Sir W.G. Armstrong Whitworth Aircraft and AV Roe & Company (Avro). Sopwith – he was knighted in 1953 – remained chairman until he retired in 1963 at the age of seventy-five. In 1977 Hawker Siddeley Group became part of the nationalised British Aerospace. Privatised in the 1980s, BAe Systems became one of the world's leading defence companies.

Harry Hawker did not live to see the success of the company bearing his name. Having survived the failure of his trans-Atlantic flight in 1919, he died when the aircraft he was due to fly in the Aerial Derby crashed at Hendon Aerodrome in 1921. He was thirty-two.

Ten years earlier he had left his native Australia to learn to fly in England. At Brooklands he had met Sopwith, who employed him as a mechanic and taught him to fly. When he was not helping to design and test Sopwith aircraft, Hawker competed in car and motor cycle races at the famous banked track. For the second time King George sent a message of condolence to Muriel Hawker. It stated: 'The nation has lost one of its most distinguished airmen.'

The designer of the Sopwith Atlantic, Wilfred George Carter, was appointed chief draughtsman of Sopwith Aviation in 1916 and chief designer of Hawker Engineering in 1920. Until he became chief designer of the Gloster Aircraft Company in 1937, Carter worked for Short Brothers, de Havilland and Avro.

Gloster was subsequently acquired by Hawker. Among the best-known aircraft with which Carter was involved were the Gauntlet and Gladiator biplanes, Britain's first jet, the Gloster-Whittle E.28/39, the twin-jet Meteor, the only Allied jet to be used operationally during the Second World War, and the delta-wing Javelin. He died in 1969 aged seventy-nine.

Sopwith Atlantic Specification

Crew	two
Length	32ft 0in (9.76m)
Wingspan	46ft 6in (14.18m)
Height	11ft 0in (3.35m)
Loaded weight	6,150lb (2,795kg)
Engine	one Rolls-Royce Eagle water-cooled V12 developing 375hp (280kN)
Speed	118mph (190kph) maximum; 105mph (169kph) cruise
Service ceiling	13,000ft (3,900m)

Chapter Four

The Big Problem

The Newfoundland weather had not improved as May gave way to June. Just like Harry Hawker two weeks earlier, John Alcock was beginning to chafe at the delay in starting his trans-Atlantic flight. His impatience was heightened by the competition posed by Admiral Kerr and the Handley Page team, which represented his closest rivals. In fact, now that Hawker and Raynham had dropped out they were virtually his only rivals.

Indeed, on the day the reassembled Vimy made its first test flight from Quidy Vidi and landed at its new base at Lester's Field, the big bomber came roaring overhead as if to taunt the Vickers crew and put them in their place. Even more galling to Alcock was the knowledge that the V/1500's pilot, Major Herbert Brackley, had been one of his students at the flying school at Brooklands before the war.

But Alcock was still confident that the Vimy could do it and he was as certain as he could be that he and Arthur Whitten Brown had the necessary skill and experience to make the flight. Both, in fact, were seasoned professional aviators.

In 1910 Alcock was an eighteen-year-old apprentice engineer in Manchester when he watched Louis Paulhan landing his Farman biplane there to complete the first flight from London and win the *Daily Mail's* £10,000 prize. His imagination had been fired; he wanted to fly.

Soon afterwards he took a job as a mechanic at Maurice Ducrocq's flying school at Brooklands. Alcock made his first solo flight in a Farman pusher biplane and in 1912 gained Aviator's Certificate number 368 issued by the Royal Aero Club. On the outbreak of war, he joined the Royal Naval Air Service and took his Farman to Eastchurch on the Isle of Sheppey, where, he wrote later, 'men and machines were badly needed'. As an instructor, his students included Reginald Warneford, who would win the Victoria Cross for shooting down the first Zeppelin over Britain.

Commissioned in 1915, Alcock was posted to Moudros on the Greek island of Lemnos. Although primarily a fighter pilot, he also flew long-distance bombing missions against Turkish targets, some of which lasted more than eight hours. On 30 September 1917, Alcock was awarded the Distinguished Service Cross for shooting down two out of three enemy aircraft threating his airfield.

Later that same day he took off in a twin-engined Handley Page O/100 bomber for a night raid on railway stations at Constantinople and Haidar Pasha. But ninety minutes from Mudros one of the engines failed. Alcock nursed the crippled aircraft for 60 miles (96km), but was unable to maintain height and had to ditch in the Gulf of Xeros. Alcock and his crew were captured by the Turks and released in November 1918.

Arthur Whitten Brown was seven years older and had a more reserved and thoughtful manner than Alcock. His knowledge of the new skill of aerial navigation had been recognised by invitations from respected journals to contribute articles on the subject. 'Teddy' Brown had been born in Glasgow to American parents and, like Alcock, he grew up in Manchester. Like his father, Brown had an interest in mechanical engineering and he served an apprenticeship with the Westinghouse Electric and Manufacturing Company.

In September 1914 Brown renounced his American citizenship to enlist in the University and Public Schools Brigade (UPS). He was commissioned in the Manchester Regiment and served on the Somme. But, he wrote later: 'I had always longed to be in the air and I obtained a transfer to the Royal Flying Corps as an observer.' He would later describe this as the second step towards his trans-Atlantic flight, the first being his enlistment in the UPS.

In November 1915 he was flying as observer to Lieutenant Henry Medlicott when their B.E.2c was hit by ground fire. Brown's left leg was shattered by a bullet and he received further serious injuries when the aircraft crashed behind enemy lines. He was a prisoner of war until 1918.

Both he and Alcock used their enforced idleness as prisoners of war to contemplate the challenges of trans-Atlantic flight. Alcock, apparently, had talked endlessly about it during his incarceration by the Turks, while Brown had the time 'to begin a careful study of the possibilities of aerial navigation'.

Alcock was demobilised from the RAF with the rank of captain on 10 March 1919. 'At last,' he wrote, 'I was free to attack the big problem of crossing the Atlantic.' Soon afterwards he was at Brooklands where Sopwith and Martinsyde had assembly plants. Both companies, he knew, had declared their interest in competing for the *Daily Mail's* Atlantic prize. At the Vickers works he was reunited with old friends including Archie Knight and Percy Maxwell Muller, who were now the company's works manager and aviation superintendent respectively.

Vickers had formed an aviation department to build airships for the Royal Navy and later aeroplanes for the Royal Flying Corps. Its F.B.5 Gunbus, of which more than 220 were built during the First World War, was essentially the world's first operational fighter. Although too late for war service, the twin-engined Vimy reflected the impressive advances in aeronautical technology made during the conflict.

That it was also capable of competing for the *Daily Mail* prize was apparent to Vickers and to Jack Alcock. 'I approached Messrs Vickers and the sporting enterprise appealed to them,' he wrote later. Quoting Archie Knight, the *Daily Sketch* later reported that the idea had been put to Alcock by Maxwell Muller, when 'the captain promptly replied: "I am on it any old time"'.

The Times would report later that the Vickers directors had been dubious about the enterprise and that it was 'only the supreme confidence of Captain Alcock, Mr Muller and Lieutenant Brown' that persuaded them to make an eleventh-hour entry. The *Sketch* reported that the arrangement with Alcock was sealed with a handshake. Until he left for Newfoundland, he haunted the Vickers factory, where his enthusiasm proved contagious, inspiring designer Reginald Pierson, the mechanics and riggers to work long hours.

In early 1919 Brown had been on the point of abandoning his ambition, having returned from three years of captivity with permanent injuries that required him to walk with a stick. He was seconded to the Ministry of Munitions in London, where he worked on aero engine design and met his future wife. He wrote later:

> When, soon after the Armistice, the ban on attempts to fly the Atlantic was lifted, I hoped that my studies of aerial navigation might be useful to one of the firms that were preparing for such a flight. Each one I approached, however, refused my proposals, and for the moment I gave up the idea.

But one day, 'entirely by chance', Brown visited the Vickers factory at Weybridge, where Maxwell Muller showed him the Vimy. 'While I was talking with the superintendent,' Brown recalled, 'Capt Alcock walked into the office. We were introduced and during the course of conversation the competition was mentioned. I then learned for the first time that Vickers were considering an entry, although not courting publicity until they should have attempted it.'

The two men hit it off immediately. The quiet reserved Brown and the stocky, ginger-haired Alcock, whose ready smile radiated confidence, began talking about the challenges of the Atlantic flight. Maxwell Muller and Alcock listened while Brown expounded his ideas on navigating aircraft on long over-water flights. 'These,' Brown wrote, 'were received favourably and the outcome of the fortunate meeting was that Vickers retained me to act as aerial navigator.'

By early May construction of the Vimy and preparations for the flight had been completed. 'This aeroplane,' *Flight* noted, 'is practically similar in every respect to the standard Vimy as supplied to His Majesty's Government.' Although some examples had been built at Vickers' works in Crayford, Kent, the one selected for the Atlantic flight was assembled at Brooklands and had the construction number 13. It would, however, make its famous flight devoid of civilian registration or other markings.

The bomb racks had been removed and the gunners' positions faired over. Fuel and oil capacity had been increased to give the aircraft a range of 2,440 miles (3,900km). One of the fuel tanks was constructed to act as a life raft. Documents in the Vickers archives at Brooklands indicate that Rolls-Royce had expressed reservations about the aircraft's fuel system, which it considered over-complex. The engine builder conceded, however, that 'it would serve' with some relatively minor amendments.

As to navigation, *Flight* reported that Brown would rely on a system similar to that employed in marine navigation. The aircraft would carry wireless instruments capable of receiving and despatching messages for a distance of 250 miles (400km) and would be able to communicate with passing vessels. Preliminary tests were carried out by Alcock and Brown and they were reported to be completely satisfied. *Flight* added: 'The Rolls-Royce engines ran perfectly and the aeroplane left the ground with its load of four tons of petrol and oil after running a very short distance on the ground.'

The Vimy was dismantled and packed in cases for shipping to Newfoundland. Apart from the two crew members, the party comprised eight men from the Vickers factory plus a Rolls-Royce engine specialist and two mechanics (*see* box). The aviators embarked on the liner *Mauretania* at Southampton while the aircraft, with the mechanics and riggers plus all the ancillary equipment, travelled separately. Arriving at Halifax, Nova Scotia, Alcock and Brown went by rail to St John's, where they 'joined the merry and hopeful company of British aviators' who had adopted the Cochrane Hotel as their base. They arrived on 13 May and were, as a result, well behind the other contestants.

The first challenge was to find a field suitable for the Vimy to start its trans-Atlantic journey weighed down by a heavy load of fuel. Newfoundland might have been the nearest part of North America to Europe, but in 1919 it was hardly ideal for the purpose. Brown wrote: 'The whole of the island has no ground that might be made into a first-class aerodrome. The district around St John's is especially difficult.'

To take off the Vimy would need a clear run into the wind of 500yd, but the two aviators could find nothing big enough. Indeed, the whole area was either wooded or rolling, hilly country with soft soil strewn with boulders and slashed by ditches.

The two men spent their first week 'driving over hundreds of miles of very bad road' in hired cars. They bought a second-hand Buick that had just 400 miles (640km) on the clock but, as Brown observed ruefully, 'Before long we were convinced that the speedometer must have been disconnected previous to the final 40,000 miles [64,000km].' They spent their evenings at the cinema in St John's or playing cards with the other competitors at the Cochrane Hotel. However, the disappearance of Hawker and Grieve created a sombre atmosphere there.

The Vimy and its accompanying team of mechanics and riggers arrived at Quidi Vidy on 26 May after the Sopwith representatives had offered use of its field to the Vickers team. Local carter Charlie Lester used teams of horses to haul the crates containing the aircraft to the field and the assembly process began almost immediately.

The mechanics worked in the open for up to fourteen hours a day. They soon discovered improvisation would be necessary. One example was the use of scaffolding poles as sheerlegs. Frederick Raynham offered the use of the Martinsyde hangar as a store and even the reporters covering the story pitched in. One provided a much-needed electric torch.

The Sopwith party also offered the use of their aerodrome, but neither it nor Quidi Vidi was considered suitable for the Vimy and the search for a suitable field continued until Charlie Lester came up with the offer of a field at Monday's Pool. It wasn't ideal but combining its useable stretch with several other adjacent fields produced enough space for a 500yd run. The original price sought by the land owner was considered extortionate, but persistent haggling brought it down to a more suitable figure.

There was still plenty of work to prepare what was now known as Lester's Field for the Vimy to use. A gang of thirty labourers worked with picks and shovels to remove hillocks, blast boulders and level walls and fences. Even then it was still not perfect, being partly uphill, but it would have to do. Despite the bad weather, the assembly work continued. The engines had to be covered in tarpaulins and windbreaks had been erected to protect the mechanics from the worst of the weather. By the morning of 9 June the aircraft was fully assembled, the engines checked and water for radiators had been filtered then boiled in a steel barrel. The aircraft was ready for its first test flight.

Despite a degree of secrecy, the news leaked out and before the engines had been warmed up a large crowd had gathered. 'The weather was on its best behaviour,' Brown recalled, 'and our take-off was perfect in every way. Under Alcock's skilful hands the Vickers Vimy became almost as nippy as a [fighter].' After take-off the aircraft headed out to sea. During its fifteen-minute flight over the ocean it passed a vivid blue surface reflecting the colour of the sky and dotted with icebergs and whitecaps.

The only problem they discovered was the wireless, from which no spark of life could be coaxed. Soon, however, smoke from a signal fire below was sighted, enabling the Vimy to head for the airfield that had been constructed for it. The landing was not without drama: the big bomber landed uphill, crested the brow and nearly ran into a fence. Alcock gunned the starboard engine and swung the Vimy over to avoid the obstacle. When the aircraft finally came to rest it was pushed down the hill and securely pegged within a roped-off space.

The Handley Page, though, had beaten them to it; Admiral Kerr's team had made their first test flight the previous day. Later, Kerr told the press: 'Everything worked most satisfactorily.' The aircraft was ready to make an attempt as soon as the weather was right. Yet the *New York Times'* man in St John's noted that 'the Vimy men are the favourites here'.

But there was another headache for the Vickers team. The mixture of petrol and benzol they had brought with them from England was found to contain sticky, resin-like deposits that dried to a powdery substance. Clearly the aviators could not risk using it on their long over-water flight. Raynham offered his spare stock of fuel, but Maxwell Muller had just arrived with enough for the journey.

A second but shorter trial flight on the 12th showed the wireless equipment as the only unsatisfactory item, giving Brown what he called 'a violent shock'. By now the aviators were 'besieging' the local RAF meteorologist, Lieutenant Lawrence Clements, for weather reports. Clements had also taken on the role of official starter on behalf of the Royal Aero Club. It was his job to fix the club's official seal to the Vimy to prove that the aircraft that took off on the Atlantic flight was the one that landed at the end of the journey.

As far as Clements could tell, the weather was now favourable for the flight and the aviators hoped to start on the 13th. Preparations were made accordingly, and the navigating equipment was loaded on to the aircraft. A coat of protective dope was applied to the aircraft's wing fabric. Alcock was keen to go. He regarded the Vimy's construction number 13 as a good augury. Maxwell Muller strongly disagreed. He felt that attempting a take-off with the wind as it was would be too risky.

Alcock was keen to ensure the preparations for the flight were as thorough as they could be. He insisted on the fuel being filtered through chamois leather and fine wire mesh. But it was during the fuelling process that there was more unexpected drama when the additional weight caused one of the aircraft's shock absorbers to break. The fuel had to be removed from the tanks before a repair could be attempted and this took all night. The mechanics worked by the light of car headlights while the flyers were sent back to the hotel at 1900 hours to get what rest they could. Dawn was breaking as the last of the 865gal (3,940lit) of fuel and 46gal (209lit) of oil went into the Vimy's tanks (*see* box).

By that time Alcock and Brown had been roused and returned to the field. Clements' latest forecast was fairly favourable. But the crosswind was still blowing and the flyers were advised to wait a few hours to see if it would abate. Their toilet kit and food for the journey – sandwiches, chocolate, Horlicks malted milk and two Ferrostat vacuum flasks of coffee – were packed on board. Emergency supplies went into a special compartment in the tail. Last to go aboard were the two mascots, soft toy cats named Twinkletoes, a present to Brown from his fiancée Kathleen Kennedy, and Lucky Jim. They went into the tail compartment. Already loaded was the first trans-Atlantic air mail, 300 personal letters each bearing special stamps.

After breakfast it was decided to take off at midday. The original plan had been to go downhill in an easterly direction. But the wind was still too strong and the team had to man-handle the Vimy to the other end of the field to enable it to take off into the wind. By 1400 hours the wind had still not dropped and a picnic lunch was eaten under the Vimy's wing. Alcock was wearing a dark blue suit, white shirt and tie topped off by a cloth cap, while Brown had donned his RAF uniform.

The wind had not dropped two hours later, but Alcock and Brown decided to go anyway. Again, Maxwell Muller tried to dissuade them: the wind was still too changeable. But they wriggled into their flying gear, electrically heated Burberry flying suits – the battery was between the seats – fur gloves and fur-lined leather helmets. Unlike Hawker and Mackenzie Grieve, they had no waterproof suits: Alcock and Brown's were still in transit.

The departure,' Brown would report, 'was quiet and undramatic.' Just before take-off he told the reporters that his plan was to head west in a straight line to the Galway coast. 'Yes,' added Alcock with a laugh, 'we shall hang our hats on the aerials of the Clifden wireless station as we go by.' After shaking hands with Clement, Maxwell Muller and other well-wishers, the two airmen climbed aboard the Vimy and squeezed themselves into a cramped cockpit dominated by the pilot's big control wheel on the right.

While Alcock ran up the engines, Brown checked his equipment: sextant clipped to the dashboard, course and distance indicator located on the side of the fuselage, drift indicator under his seat, and Baker navigation machine containing the necessary charts on the cockpit floor. Brown had also packed a torch and there was a Very signal pistol with red and white flares within easy reach. Last of all, Maxwell Muller handed up Brown's walking stick.

The latest weather forecast showed that the strong westerly wind would drop before the aircraft was 100 miles (160km) into its trans-oceanic journey. Wind velocity was predicted not to exceed 23mph (37kph) and clear weather was expected for most of the trip. But, as Brown would recall ruefully, 'The promised clear weather never happened.'

The uphill take-off proved no easier than expected. Brown wrote: 'With throttles open and engines all out the Vickers Vimy lumbered into the westerly wind.' The wind was still gusting at up to 7mph (12kph) and the runway surface was bumpy. The heavily loaded aircraft bounced along for more than 300yd before staggering into the air.

Once or twice its wheels nearly touched the ground again. 'It's down!' went the cry from the onlookers. It was not, but during its laborious climb, Brown was worried that the aircraft might hit a roof or a treetop. He noticed the sweat running down Alcock's face as he tried to coax the Vimy into the air. Getting it to 1,000ft took eight minutes. When it crossed the coast, Brown unwound the wireless aerial and tapped out the message: 'All well and started.' As he did so he thought: 'Behind and below is America, far ahead and below is Europe.' Maxwell Muller sent a telegram to Weybridge: 'Vimy left here one-thirty today [UK time].'

At first good progress was made; a tailwind was pushing them along at nearly 16mph (258kph). For the first hour visibility was good and the sky was visible through gaps in the clouds. Below, the sea gleamed blue-grey dotted with white icebergs. Kneeling on his seat, Brown used the Baker navigating machine to take observations of the sea, the horizon and the sun.

Brown had already resolved to use such observations combined with dead reckoning rather than relying on the directional wireless equipment. This was to be a sensible precaution. He had an ordinary marine sextant with a more heavily engraved scale than normal to make it easier to read in the constantly vibrating aircraft. He had several charts, the main one based on the Mercator projection with another for use at night. To measure drift he had a 6in (15cm) drift bearing plate that enabled him to measure ground speed with the help of a stopwatch. As he expected the horizon to be obscured by cloud or mist for much of the time, Brown had an instrument like a spirit level in which the horizon was represented by a bubble. Brown wrote:

> I could congratulate myself legitimately on having collected as many early observations as possible, while the conditions were yet good for soon we ran into an immense bank of fog which shut off completely the surface of the ocean. The blue of the sea merged into a hazy purple and then into the dullest kind of grey.

The cloud screen above thickened and there were no more gaps. The occasional glints of the sun on the Vimy's wing tips and struts no longer appeared. It was not possible to take observations on the sun nor to calculate the drift from the sea. Brown had to rely on his earlier observations, proceed by dead reckoning and 'hope for the best'.

At 17:20 hours the aircraft was at 1,500ft and still climbing, but the haze was growing thicker and heavier. Brown attempted to send a message using the wireless transmitter, but the small propeller that drove the generator suddenly snapped and broke away. He could still receive messages, but not transmit. This silence from the Vimy would worry the rest of the team awaiting news in Newfoundland.

Another item of communications technology that would be rendered useless was the intercom device. Tiring of the discomfort they caused, Alcock ripped the earpieces from his helmet. Henceforth the aviators would communicate by signs and scribbled notes. In the first of these, Brown passed revised course directions. In the second he told the pilot of the wireless failure: 'Wireless generator smashed. The propeller has gone.' For the rest of the evening the aircraft flew on through the thick fog that completely cut off the aviators' view of the sea. 'I can't get an obs in this fog,' Brown told Alcock in another scribbled message. 'Will estimate that same wind holds good and work by dead reckoning.'

But, Brown noted dryly: 'The early evening was by no means dull.' Just after 1800 hours the starboard engine startled the aviators by departing from its normal roar with a noise 'like machine gun fire at close quarters'. For a second or two the aviators feared the worst: were they about to suffer the same fate as Hawker and Grieve?

They did not have to look too hard at the engine to see what had happened. A chunk of exhaust pipe had split away and was 'quivering before the rush of air like a reed in an organ pipe'. The section of piping became red hot, then white hot, before gradually crumpling up and blowing away in the slipstream. Three of the twelve cylinders were now exhausting directly into the air. It was, said Brown, 'a minor disaster, unpleasant but irremediable'. Their ears soon became accustomed to the noise, but Brown noticed that flame was flicking from the open exhaust directly on to a bracing wire, making it red hot.

At about 1900 hours the aircraft emerged into clearer air, although there was another layer of cloud above it. Half an hour later Brown handed Alcock two sandwiches and some chocolate. He uncorked the vacuum flask and handed it to the pilot, who would only take one hand off the controls to eat and drink. About an hour later a sizeable gap in the clouds above enabled Brown to take an observation of the sun. His subsequent calculations confirmed that they were on course. At 21:15 hours the sea became visible briefly, enabling Brown to tell Alcock that they were still travelling at 160mph (256kph) but that they were probably further east and south than expected.

By now the Vimy was travelling at 4,000ft (1,220m) with dense cloud above and below. Alcock continued to climb in the hope of getting above the murk. At 22:00 hours Brown had to switch on the compass light. He now had to use his torch for his periodic inspection of the engines. By 23:20 they were flying at 5,200ft, but observations were impossible and by midnight they had reached 6,000ft. Patchy cloud meant that some star shots were possible and Brown was able to establish the Vimy's position: they had already covered 978 miles (1,565km) and were slightly to the south of their correct course. The reason, he thought, was that his calculations had been influenced by the forecasts received at St John's.

To nurse the engines Alcock allowed the altitude to fall off, so that by 02:20 hours the aircraft was back down below 4,000ft (1,200m). The fog was back too, making drift readings virtually impossible. The moon was partially visible, but the horizon was indistinct. Brown was struck by the 'fantastic surroundings', which he called 'extravagantly abnormal'. Later he would describe 'the distorted ball of a moon, the eerie half-light, the monstrous cloud shapes, the fog below and around us the misty indefiniteness of space, the changeless drone, drone, drone of the engines'.

His response was to turn to refreshment. 'Twice during the night we drank and ate in snatches, Alcock keeping a hand on the joystick, while using his other to take the sandwiches, chocolate and thermos flask, which I passed to him one at a time.' By now they were beginning to feel the constraints of the cramped cockpit, but Brown said that neither he nor Alcock had felt sleepy.

A clear patch in the clouds enabled Brown to fix the Vimy's position from the Pole Star, Vega and the moon. As they began to think about the new day and what it might bring things seemed to be going reasonably well. Thanks to Alcock's careful management of the flight, keeping engine power between 50 to 75 per cent throttle, the Vimy had plenty of fuel left in its tanks. Their course, too, seemed satisfactory. Perhaps it was too good to be true.

At about sunrise the aircraft suddenly ran into a thick bank of fog that projected below the lower cloud layer. 'Separated suddenly from external guidance,' Brown would report, 'we lost our instinct of balance. The machine, left to its own devices, swung, flew amok and began to perform circus tricks.'

Then came what Alcock would later describe as 'a mishap that nearly brought our venture to an untimely end'. The airspeed indicator suddenly failed to register. Turbulence prevented him from holding a course. The Vimy was rocking from side to side and the crew began to lose their equilibrium.

In his post-flight interview with a *Daily Mail* reporter, Alcock said: 'We looped the loop, I do believe, and did a very steep spiral. We did some very comic stunts for I have had no sense of horizon.'

Brown recalled that the Vimy had briefly hung motionless as it teetered on the brink of a stall. Then it heeled over and twirled rapidly into a spin. The compass needle went mad and the thick fog made it impossible for the aviators to determine the direction of the spin. The aircraft vibrated as the engine revs rose. Alcock throttled back and the vibration stopped.

Alcock wrote later:

> From an altitude of 4,000ft we twirled rapidly downward. It was a tense moment for us and when at last we emerged from the fog we were close down over the water at an extremely dangerous angle. The white capped waves were rolling along too close to be comfortable.

As the Vimy spiralled down, Brown prepared for the worst, loosening his seat belt and preparing to save his log of the flight. He was well aware that their chances of surviving a crash into the sea were slim.

But the murk cleared just enough for a quick glance at the horizon. This enabled Alcock to regain control and get the aircraft back on course. Brown reckoned they were down to within 100ft of the sea, and by the time Alcock had regained control this had shrunk to just 50ft with the waves almost within touching distance. As the aircraft levelled off it seemed as if the surface was level with the aircraft and sideways on to it.

By now the aircraft was facing the wrong way and Alcock had to turn the Vimy in a wide semi-circle as it climbed to regain height. But there was still no break in the cloud; for the next three hours the aircraft would be totally enveloped. The aviators' world shrank to a few yards from each wing tip.

And with the dawn came more bad weather. Heavy rain and snow turned to hail and sleet. Although open to the elements, the Vimy's cockpit kept the crew relatively snug providing they remained in their seats. A hand exposed above the coaming would receive stinging stabs from the hailstones that were now lashing down.

Then came the event that would go down in legend as the most dramatic moment of the flight. It was Brown's job to monitor the gauges mounted on the struts behind the cockpit, particularly the petrol overflow gauge. This was critical to enable the pilot to ensure that the engines were supplied with fuel at consistent pressure. Failure to do so could mean them becoming starved of fuel.

When the aircraft had climbed to 8,800ft (2,680m) Brown noticed that the glass face of the gauge had become obscured by snow. It was up to him to clear it. 'The only way to reach it,' he said later, 'was by climbing out of the cockpit and kneeling on top of the fuselage while holding a strut for the maintenance of balance.'

In other accounts Brown said that he had been able to clear the face of the gauge by kneeling on his seat and reaching up. There was no mention of kneeling on the fuselage. But however it happened, it must have been an effort for a man bundled up in heavy clothing and hampered by an injured leg. And the sudden change from the comparative warmth of the cockpit to the biting cold slipstream came as a shock. 'Startlingly unpleasant', was how he would later describe it. 'The violent rush of air which tended to push me backward was another discomfort.'

In his account of the journey published in the *Royal Air Force and Civil Aviation Record*, Brown recalled: 'I had no difficulty, however, in reaching upward and rubbing the snow from the face of the gauge.' Until the storm ended, a repetition of the performance, at fairly frequent intervals, continued to be necessary. Kneeling on the fuselage presented little danger of falling providing Alcock kept the machine level.

But in subsequent years some accounts would maintain that he had hauled himself out of the cockpit and inched along the wings to hack away with his knife the ice that threatened to choke the engines and hurl the Vimy and its crew into the sea. One report claimed he had repeated this feat six times.

Alcock's first account to the *Daily Mail* and featured in the newspaper the day after the Vimy's arrival in Ireland made no mention of it, even though he started it with the words: 'We have had a terrible journey.' For four hours, he said, the aircraft was covered in a sheet of ice caused by frozen sleet, which at one time was so dense that the air speed indicator did not work.

But in an account published later Alcock was quoted as saying: 'My radiator shutter and the water temperature indicator were covered with ice for four or five hours. Brown had continually to climb up to chip off the ice with a knife. The air speed indicator was full of frozen particles and gave trouble again. They came out when we got lower an hour before we landed.'

In an article published in 1969 to commemorate the flight's fiftieth anniversary, the journal *Flight* recalled: 'Brown ... had to climb out first on to one wing, then on to the other to clear the [engine air] intakes by hand. Handicapped by his injured leg he clambered along the slippery wings in full force of the icy slipstream and cleared the intakes with his pocket knife.'

In his contribution to *Our Transatlantic Flight*, also published in 1969, Alcock's brother, Edward, also known as John, added to the myth when he wrote: 'My brother told me afterwards that if Brown had not climbed out on the wing to clear the ice they would never have made it.'

With the Vimy's airspeed indicator and engine revolution counters mounted outside the cockpit, Captain John Alcock said that without them his brother would not have been able to keep the aircraft flying. He wrote: 'It was these that Brown had to clear if the 'plane was not going to dive into the Atlantic or go into such a steep climb that she would stall. Added to their problems, the ice was blocking the air intakes to the carburettors; these too Brown had to clear.'

Anybody who has seen the Vimy, either the real thing in the Science Museum in London or the replica at Brooklands, can hardly fail to be impressed by the extreme difficulty – not to say impossibility – of such a manoeuvre in the face of the fierce slipstream, especially for a man with a damaged leg.

Yet the legend has persisted. The *Flight* account was typical of the stories that appeared after Brown's death. So were they simply the result of misunderstanding or a somewhat liberal interpretation of the aviators' published accounts?

Back in the Vimy the snow and sleet was still falling. The aircraft, though, was still climbing so that by 05:00 it was at 11,000ft (3,350m). A clear patch enabled readings to be taken that indicated the Vimy was nearing the Irish coast. But the weather had not finished with them. At 06:20 Brown noted in his log that the upper wing surfaces were covered in frozen sleet. And the sleet was also jamming the aileron hinges, virtually robbing Alcock of any lateral control. An hour later Brown passed Alcock a note asking him to go down lower to warmer air.

As the nose dipped the aviators were horrified to hear the starboard engine begin to pop ominously. Alcock throttled back and the noise stopped. By 08:00 they were down to 1,000ft. The air was warmer and the ailerons were functioning again, but they were still surrounded by mist and murk hiding the surface of the sea. As it was not certain that the altimeter could be relied upon, due to possible changes in barometric pressure, there were fresh fears of hitting the water.

'Once again we were lucky,' Brown wrote. 'At a height of 500ft [150m] the Vickers Vimy emerged from the pall of cloud and we saw the ocean – a restless surface of dull grey.' Alcock opened the throttles and the two Rolls-Royce Eagles responded. Brown's calculations now put them slightly to the north of their planned route. Passing a revised course of 170 degrees, he told Alcock: 'Don't be afraid of going S. We have had too much N already.'

The Vickers Vimy

Although conceived as a bomber, the Vimy was to act as a harbinger of peacetime civil aviation through some spectacular long-distance flights.

Vickers' first twin-engined bomber was conceived and designed by chief designer Reginald Kirshaw 'Rex' Pierson within four months to meet an Air Board requirement for a heavy bomber. The prototype (B9952) flew for the first time at Joyce Green near Dartford on 30 November 1917.

The aircraft was of conventional construction with a fuselage built of steel tubing and covered in plywood and fabric. The biplane wings were built around two spars with their shape defined by closely spaced wooden ribs. The wings and biplane tail were fabric covered and the whole structure was braced by a system of struts and wires.

Four prototypes were built and tested with Hispano-Suiza, Salmson, Sunbeam, Fiat and Rolls-Royce engines, although it was the latter's Eagle VIII units that were fitted to production FB.27A aircraft. Rolls-Royce had been developing the Eagle (and the smaller Falcon) since 1914. Impressed by the success of the engine used by the Mercedes car that had dominated that year's French Grand Prix, Royce determined that the units should be liquid-cooled with the cylinders arranged in a vee.

But it was with Hispano power that the first FB.27 went in January 1918 to the Aeroplane Experimental Establishment at Martlesham Heath for official trials. It caused a sensation by lifting a payload heavier than its main competitor, the Handley Page O/400, which had twice the power. However, engine problems meant that the Vimy had to return to Joyce Green, where a number of alternative power plants were tried.

As the Vimy had the range to reach Berlin from bases in France it had been intended to deploy the aircraft to France for long-range bombing missions over Germany, but by October 1918 only three had been delivered. It was not until July 1919 that the type went into service with the RAF's No. 58 Sqn in Egypt. During the 1920s Vimys formed the RAF's main heavy bomber force and represented the only twin-engine bombers stationed at home bases.

More than 1,000 Vimys had been ordered, but most were cancelled after the Armistice. Vickers built 147 aircraft at Bexleyheath, Crayford and Weybridge, and further examples were to have completed by a number of other companies. In the confusion of cancelled orders and unfinished aircraft the number actually built remains uncertain, but is probably more than 230.

Had the First World War continued Vimys might well have played an important role in the conflict. But it was exploits such as Alcock and Brown's first non-stop trans-Atlantic flight that won undying fame for the aircraft. Ross and Keith Smith, together with sergeants W.H. Shiers and J.M. Bennett, made the first flight to Australia in just under twenty-eight days in November and December 1919.

Less successful were the attempts to reach South Africa. Lieutenant Colonel Pierre Van Ryneveld and Major Quintin Brand set out from Brooklands in February 1920 in a Vimy but reached Cape Town in a borrowed DH.9 after a series of mishaps that saw them changing to a second Vimy before abandoning it. Like the Smith brothers, Van Ryneveld and Brand were knighted. Another Vimy, chartered by *The Times*, also crashed en route to South Africa in 1920.

Vickers Vimy Specification

Length	43ft 7in (13.28m)
Wingspan	68ft 1in (20.75m)
Height	15ft 8in (4.77m)
Maximum take-off weight	10,884lb (4,937kg)
Power plant	two Rolls-Royce Eagle V12 engines each developing 360hp (268kW)
Max speed	100mph (161kph)
Range	900 miles (1,448km)
Service ceiling	7,000ft (2,134m)
Armament	2,476lb (1,123kg) bombload plus two 0.303in (7.7mm) Lewis machine guns on Scarff mountings in nose and mid-fuselage.

Chapter Five

What do you Think of that for Fancy Navigating?

By 08:00 hours on 15 June Arthur Whitten Brown was pretty confident that, if his calculations were correct, the Vickers Vimy was nearing the end of its journey. He and John Alcock were on the verge of completing the first-ever non-stop aerial crossing of the Atlantic.

'Although neither of us felt hungry,' Brown would write later, 'we decided to breakfast at eight o'clock, partly to kill time and partly to take our minds from the rising excitement induced by the hope that we might sight land at any instant.'

While the aircraft was flying at 200 to 300ft (60 to 90m), Brown placed a sandwich and a piece of chocolate in Alcock's left hand. The pilot's right hand and both his feet remained on the controls. He had maintained this position without respite for the past sixteen hours. As the fuel was consumed the aircraft adopted a nose-heavy attitude, compelling Alcock to maintain a constant backward pressure on the control column.

Brown's expectations of sighting land were soon to be justified. He was just replacing the thermos flask and remnants of breakfast in the cockpit cupboard when Alcock grabbed his shoulder, twisted him around and pointed ahead and below. Brown wrote later: 'His lips were moving, but whatever he said was inaudible above the roar of the engines. I followed the direction indicated by his outstretched forefinger.'

Barely visible in the mist were two specks of green. Land! They had sighted the tiny Eeshal and Turbot, two of a group of islands just off the Galway coast. 'We were jolly pleased, I tell you,' Alcock told reporters later. They crossed the Irish coast at 08:25 hours. The Vimy then circled the small town of Clifden before Alcock spotted the tall masts of the Marconi wireless station at Derrygimla, south of the town.

It was Sunday morning and many of the town's inhabitants were in church. Some saw the Vimy and most heard the roar of its engines. The aircraft headed towards the station and circled the town, firing Very signals. Alcock wrote: 'By the wireless station I had noticed what I took to be a suitable field so we decided to land there. But what I thought was a field turned out to be a bog.'

When it touched down the aircraft ran for about 50yd (46m) before the wheels began to sink into the soft ground. The aircraft stopped suddenly and as the front landing skid had been removed there was nothing to stop the nose burying itself in the boggy soil. It was 08:40 hours on Sunday, 15 June. Little more than sixteen hours earlier Alcock and Brown had left Newfoundland.

Almost immediately, men from the military detachment at Clifden hurried to the scene. There are several accounts of what happened next. According to Brown's the two fliers were asked: 'Anybody hurt?'

'No,' was the aviators' reply.

As they were helped from the cockpit they were asked: 'Where are you from?'

The reply brought polite laughter and the response: 'America?' According to one report, the pilot said: 'I'm Alcock, just come from Newfoundland.' Brown was said to have commented: 'That's the way to fly the Atlantic!'

Once the truth had sunk in, the two flyers were asked for their autographs. It would be the first of many times they would sign their names over the next few days. 'Now,' Alcock was reported as saying, 'if we had a shave and bath, we should be all right.'

The two were led away to the wireless station, walking stiffly after their confinement in the Vimy's tiny cockpit and burdened by their heavy flying kit. The journey over the boggy ground was, Brown recalled, 'a dragging discomfort'. The noise of the engines had made them temporarily deaf. They were, however, revived somewhat by their reception in the officers' mess.

Later they were interviewed by the reporters who had been awaiting their arrival in Ireland. Although, the damage to the Vimy was relatively slight. it was clear that it was not in a fit state for the planned onward flight to Brooklands. 'The only thing that upset me,' Alcock said, 'was to see the machine at the end get damaged.'

The *Daily Mail* reported that, on landing, Brown had said to Alcock: 'What do you think of that for fancy navigating?' 'Very good,' was the reply and they both shook hands.' The paper's correspondent on the spot had been one of the first newsmen to congratulate the airmen on their success. They were in the officers' mess at the wireless station packing up their flying gear and navigation equipment as the reporter recorded their reactions for posterity:

> Capt Alcock's big ruddy face just lit into a smile and Lt Brown bent further over his sextant and said simply, 'We didn't do badly, did we?' They might have come ten miles. Alcock told him with a laugh that he was not at all tired but Lt Brown confessed, 'I am a bit fagged out.'

The following morning Alcock and Brown awoke in Ireland, having slept for nine hours after their exhausting flight from Newfoundland. They felt refreshed but, as Brown would write later, 'strange'. He recalled: 'Our physical systems, having accustomed themselves to habits regulated by the clocks of Newfoundland, we were reluctant to rise at 7 am for subconsciousness suggested it was but 3.30 am.' Alcock and Brown might have been the first trans-Atlantic flyers, but they would certainly not be the last to feel the effects of travelling across time zones. Later generations would talk about jet lag.

Over breakfast the two aviators were photographed reading accounts of their flight in the morning's newspapers. They were starting their first full day as celebrities: on their journey back to London the pair would be greeted by large and enthusiastic crowds. Flowers would be offered and autographs sought.

The journey to Dublin gave them a taste of what awaited them in England. 'We must have signed our names hundreds of times,' Brown recalled, 'on books, cards, old envelopes and scraps of paper of every shape and every state of cleanliness.'

At Holyhead they were welcomed by Reginald 'Rex' Pierson, designer of the Vimy, and other representatives of Vickers and Rolls-Royce. Crowds were waiting to catch sight of the aviators all the way to London. At Chester, Crewe, Rugby and other towns they were welcomed by mayors and civic heads. If anything, the crowds were even more enthusiastic in London and they were mobbed at Euston. From there they were driven to the Royal Aero Club, where Alcock delivered the bag of mail he had been handed in Newfoundland. The two aviators made speeches and later appeared on the balcony to more cheers from the crowd gathered below.

That evening Alcock went to watch Joe Beckett knock out Frank Goddard for the British heavyweight title at Olympia. Brown went to visit his fiancée, Kathleen Kennedy. Soon after landing he had cabled her: 'Landed Clifden Ireland safely this morning. Will be with you very soon. Teddy.'

The reply had come by early afternoon: 'Magnificent – never doubted your success. Wire when leaving for Brooklands. Will meet you there. Micki.'

In fact, the couple had been reunited when the train carrying the aviators to London stopped at Rugby.

The following day Alcock and Brown arrived at Brooklands. It had not been quite the return to Vickers' base that Alcock had hoped for, but the aviators received what Brown called 'the welcome we appreciated most'. He added: 'We were chaired and cheered by the men and girls who had built our trans-Atlantic craft.' Vickers' employees at Brooklands had been given the day off to celebrate, as had their counterparts at Rolls-Royce. *The Daily Chronicle* reported that news of the flight had 'spread like wildfire and Vickers' offices were inundated with inquiries and congratulations'.

On the 20th the aviators were guests of honour at a luncheon at the Savoy hosted by the *Daily Mail*. Lord Northcliffe, the man who had offered the £10,000 prize for the first trans-Atlantic crossing, was not able to be there because of illness. But Thomas Marlowe, chairman of Associated Newspapers and editor of the *Mail*, saluted Alcock and Brown's achievement as 'one of the greatest in the history of human progress'.

Handing over the newspaper's cheque, together with additional prizes totalling £3,100, Winston Churchill, Secretary of State for Air, said it was 'an achievement which marks the advance of science and of engineering and the increasing triumph of men over nature'. He added: 'No one has ever flown the Atlantic in a single bound before and no one will ever fly the Atlantic for the first time in a single bound again.' He pointed out that the flight had demonstrated aviation's rapid progress since it was in 1909 that Louis Bleriot had flown the English Channel; now, just ten years later, Alcock and Brown had crossed the Atlantic.

The following day the aviators went to Windsor to be knighted by the King. Afterwards, *The Observer* reported that the aviators had been driven from the station to the castle by a royal carriage 'drawn by a pair of greys'. They 'received a great ovation from the Eton boys who surrounded the carriage and ran with it to the castle, waving their top hats and cheering'. The newspaper reported that, after the King and other members of the Royal Family had enjoyed an 'interesting chat' with the aviators, the crowd waiting at the station was so big that 'it was with the greatest difficulty that the officials got them into the train'.

Flight called the knighthood bestowed on the flyers 'fitting' and compared them with Sir Francis Drake who, centuries earlier, had been the first Englishman to circumnavigate the globe. 'Alcock and Brown went off into the unexplored air fired with intent to bring home to Britain the honour of the first direct flight across the Atlantic,' the journal declared. 'As long as history exists theirs will be the names associated with this great achievement.'

The Times, however, struck a discordant note when it revealed that some experts had questioned Brown's professional skills. It reported: 'Lt Brown, the navigator, is a landsman who has confounded the sailor critics who declared that the Atlantic could be navigated only by a blue water sailor.'

But such negative comments were overwhelmed by the euphoria that followed the return home of the two aviators: two Britons in a British aircraft had beaten the world to win a big cash prize. Inevitably there was much talk of what their flight portended. Winston Churchill observed that the flight's main significance lay in drawing together 'the great English-speaking communities that dwell on both sides of the Atlantic'.

Press baron Lord Northcliffe, who had put up the prize money for the flight, chose to see the Atlantic crossing in terms of its significance to his own business. He told Alcock and Brown that their achievement had sent a strong message to what he called the 'cable monopolists' who would have to 'improve their wires and speed up'. It had taken Alcock and Brown less time to cross the Atlantic 'than the average press message of 1919'. He also looked forward to the days when 'London newspapers will be selling in New York in the evening, allowing for the difference between English and American time'.

Yet amid the back-slapping and speeches there was a feeling of anti-climax. The Atlantic had been flown and the *Daily Mail* prize claimed, so what would come next? The *Manchester Guardian*, although full of praise for the aviators and their achievement, reflected on its meaning:

> The success of Capt Alcock and Lt Brown, while proving not for the first time that they are very gallant men does not, it must be confessed, prove much else ... As far as can be foreseen, the future of air transport over the Atlantic is not for the aeroplane. It may be used many times for personal feats of daring. But to make the aeroplane safe enough for business use on such sea routes we should have to have all the cyclones of the Atlantic marked on the chart and their progress marked in from hour to hour ...

The Vimy's flight had certainly been a graphic demonstration of the aeroplane's rapid development, but there were many who doubted that it really represented a pointer to the future. In the summer of 1919 the immediate future of long-distance oceanic travel seemed to belong to lighter-than-air craft.

Brown, the man who had navigated the Vimy across the Atlantic with such precision, was firmly convinced that no aeroplane yet built or contemplated had the ability to fly non-stop between London and New York. Even with intermediate halts, aeroplanes would only be able to carry mail or low-weight items.

Despite his and Alcock's success, Brown believed that 'it is folly to expect an air age now'. He said: 'Its coming will be delayed by the necessity of slow, painstaking research and by the fact that in the countries which are encouraging aviation to the greatest degree, capital is no longer fluid and plentiful and money in substantial sums cannot be risked on magnificent experiments.'

He was right that commercial trans-Atlantic aviation would not arrive for many years to come. Aviation technology was nothing like ready for commercial operations. Indeed, it would be another quarter of a century before regular trans-Atlantic passenger flights became reality. And not until 1958 did the number of passengers travelling by air exceed those going by sea.

By the time of Charles Lindbergh's non-stop solo flight from New York to Paris in 1927 fewer than 120 people had crossed the Atlantic by air and more than half of them had done so aboard a single vessel, the British airship *R34*.

It was clear that the amount of fuel and lubricating oil carried by the Vimy represented more than 40 per cent of its total take-off weight (*see* box). And it did not take a mathematical genius to work out that once the weight of the crew and their equipment had been included there was precious little scope for any worthwhile payload to be carried.

Once the *Daily Mail* prize had been claimed the reporters covering the story from Newfoundland moved on and Admiral Kerr and the crew of the big Handley Page bomber decided to do the same. At one time, Alcock and Brown had seen the V/1500 as their biggest rival and it was to forestall them that Alcock had insisted on making the attempt before the wind had dropped in Newfoundland. What he did not know was that the big bomber had already encountered the trouble that would effectively remove it from contention. A defective cooling system would delay its departure until new, larger radiators had arrived from England. As a result, Kerr and his crew decided to fly south to the USA.

On the way to New York engine trouble forced them to land at Parrsboro, Nova Scotia, on 4 July. The rough landing damaged the aircraft so badly that it took nearly four months and a new fuselage shipped over from England to get it airworthy again. The flight continued to New York on 9 October.

While waiting at Newfoundland, Alcock had declared that even if the V/1500 made the first non-stop crossing the Vickers team would still make the flight to demonstrate the Vimy's capability. Now, after his triumphant return to London, Alcock was revealing in an interview with the London *Evening News* that he and Brown had decided to give £2,000 of their winnings to the men who had built the Vimy. He also said that he had turned down offers to go to work in America and had decided to remain with Vickers as its test and experimental pilot.

Sadly, he was not given long to savour the fame his achievement had brought him. On 18 December, six months after his trans-Atlantic triumph, Alcock was flying a Vickers Viking amphibian to an air show in Paris. On the way he became lost in fog and was fatally injured while attempting a forced landing in mist near Rouen.

Brown was profoundly upset by Alcock's death, but he was to live long enough to see another war and also the beginning of regular trans-Atlantic commercial services. After his flight in the Vimy, Brown returned to work for British Westinghouse, which had been renamed Metropolitan-Vickers (Metro-Vick). In the early years of the Second World War he served as a lieutenant colonel in the Home Guard before re-joining the RAF to work on navigation training. Declining health forced him to resign in 1943. The death a year later of his only son, an RAF Mosquito pilot, while on a bombing raid caused it to deteriorate further.

Sir Arthur Whitten Brown, for nearly thirty years the sole survivor of the first non-stop flight from North America to Europe, died in 1948. His widow, Kathleen, whom he married soon after his return from Clifden, survived him by only four years. But what he called his 'fancy navigation', would, however, ensure he lives on in history.

Trans-Atlantic Details

Two documents from the Vickers archives cast an interesting light on some of the details of the trans-Atlantic flight.

Personnel and Equipment taken to Newfoundland for the flight

Personnel
1 pilot
1 navigator
1 foreman in charge
1 erector
2 riggers
4 carpenters
1 Rolls-Royce representative
2 engine mechanics
All took sleeping bags, personal luggage and appropriate tool kits

Machine Spares
1 chassis complete
1 radiator
1 engine
1 tail skid
1 main plane (not assembled)
1 set RAF wires
1 set piping complete (oil, petrol and water)
1 set fabric covers for planes
2 petrol pumps
4 windmills for petrol pumps
1 pair tail plane skids

Handling and Housing
1 Hervieu hangar 80ft × 70ft (afterwards sold in Newfoundland)
2 sets blocks, falls, slings, ropes, etc
2 rotary petrol pumps for filling machine tanks
1 seven-seater Buick car was obtained in Newfoundland and afterwards sold

Chocks and trestles were made on the spot from packing case material
Miscellaneous
Blow lamps, tape, string, bench vices, dope, solder, paint, varnish
1,948 gallons of fuel (petrol and benzole)
80 gallons Castrol

Total weight of Vimy on departure from Newfoundland

	lb	kg
Weight of aircraft empty except for water	7,350	3,334
Petrol (865gal, 3,933lit)	6,230	2,826
Oil (46gal, 209lit)	460	209
Reserve water (9gal, 41lit)	90	41
Crew	360	163
Sundries	200	91
Total	14,690	6,663

Chapter Six

Another Milestone

But for a delay caused by careless handling after a test flight, Alcock and Brown might have been banished to a footnote of history by another crew in a very different craft to the Vickers Vimy.

As it is, though, history seems to have bypassed the story of George Scott, Edward Maitland and the crew of His Majesty's Airship *R34*. Yet in July 1919, barely a month after Alcock and Brown's triumph, *R34* not only made the first east-west trans-Atlantic flight, but followed it up with the first double crossing.

To twenty-first century readers the story of this epic voyage has a whiff of Jules Verne or H.G. Wells about it. Yet in 1919 there were many in aviation who believed that it was *R34* rather than the Vimy that signposted the future of intercontinental air travel. In fact, from the earliest days of powered flight there had been two main strands of development – heavier than air and lighter than air. For long-distance work the latter was considered the better with its superior range and load-carrying ability, even though the aeroplane was obviously faster.

Flight considered that the *R34*'s accomplishment marked 'the first really successful attempt to demonstrate that the airship is, so far as human vision can foresee, the real aerial vehicle for sustained flight'. The flight had also shown that such craft were 'capable of being navigated with the precision and certainty of a steamship'.

Another who believed in the airship's potential was none other than Sir Arthur Whitten Brown, the man who had navigated the Vimy across the Atlantic. In his contribution to *Our Transatlantic Flight,* which was first published in 1920 in instalments in the *Royal Air Force and Civil Aviation Record*, he observed that airships seemed to offer almost unlimited possibilities. 'A large airship,' he wrote, 'can easily provide first-rate living, sleeping and dining quarters besides room for passengers to take exercise by walking along the length of the inside keel or on a shelter deck.'

As had been demonstrated, airships could fly from London to New York in only four hours longer than an aeroplane going via Ireland and Newfoundland. Besides, Brown noted, the economics favoured the airship over the aeroplane.

The *R34* story began, curiously enough, with the attacks on Britain made by German Zeppelins in the First World War. German airship technology was tacitly acknowledged to represent the current start of the art and when the opportunity arose to examine an example that had been captured virtually intact, despite the best efforts of its crew, it was gratefully accepted. Its secrets could now be uncovered and aspects of German airship design incorporated in the design of the new patrol craft ordered by the Royal Naval Air Service in 1916.

There were two of them, designated *R33* and *R34*. The former was built by Armstrong Whitworth in Yorkshire and the latter by William Beardmore and Co. of Inchinnan, Renfrew, near Glasgow. Assembly began in 1918.

R33 was the first to fly, in March 1919. After a fairly unremarkable career, which included experimental work with 'parasite' fighters, it was broken up in 1928. *R34*'s maiden outing came less than a fortnight after *R33*'s and was followed ten days later by an extended trial flight. The weather was bad and the craft sustained minor damage, but on its return to Inchinnan poor handling by the ground crew resulted in further damage that necessitated lengthy repairs.

On 28 May the airship left Inchinnan for the 80-mile (128km) delivery flight to East Fortune. It turned into a twenty-one-hour odyssey. Thick fog made navigation difficult and forced the airship to spend the night over the North Sea. There was no food on board, so the crew had to go hungry. The fog was still there next morning and the crew was sent south, reaching Yorkshire before the weather improved.

In addition to their extensive reports of Alcock and Brown's reception by London crowds, the newspapers reported that *R34* had made a 'final' trial flight from East Fortune, primarily to test its wireless apparatus. This, *The Times* noted, consisted of three sets of high-power auxiliary and wireless telegraphy: the high-power wireless had a radius of 1,500 miles (2,400km) and the wireless telephone a speaking range of 100 miles (160km).

R34 was accompanied on the trial flight by another airship: *R29* left at 10:00 hours and *R34* thirty minutes later. *R34* required a handling party of 400 RAF men and WRAF girls to ease the airship out of its shed. On board was General Groves, deputy chief of staff at the Air Ministry, who arrived by air from London just in time to go aboard.

The craft headed north-east and then west, at an average height of 1,500ft (450m). Wireless communication was established with the Azores, 1,200 miles (1,920km) away. By 02:00 hours the weather had changed and the airships were ordered to return to their base. The wind had risen slightly and was now, 'coming in gusts across the landing ground and making the conditions somewhat awkward', according to *The Times*. An excellent landing was, however, made. 'The airship was almost beside her shed and the only mishap of the trial took place. A strong gust of wind caused the ship to bump. There was a crash and the rear car grounded somewhat heavily and some damage was done to the framework.'

Nevertheless, this augured well for a trans-Atlantic flight and the Air Ministry decided to attempt the first return crossing. But as *R34* had not been intended as a passenger carrier, extra accommodation had to be arranged with hammocks slung in the keel walkway. A plate was welded to an engine exhaust pipe to allow for the preparation of hot food. Food lockers replaced bomb racks and a compass was installed on the upper gun platform to avoid interference of the magnetic field by electrical equipment. Wash basins were added to the crew space and more tables were provided. Lightweight curtains were hung to stop draughts coming from the interior of the hull. Along the keel an additional twenty-four petrol tanks were fitted to take total fuel capacity to around 6,000gal (27,280lit).

A northerly coastal route was selected for the outward journey. If the craft ran out of fuel it would never be too far from land. Two warships, the battlecruisers *Renown* and *Tiger*, were stationed along the route to pass meteorological reports and, if necessary, to act as rescue ships. At the Admiralty a room was set aside for communication with the airship. A map was also provided to chart its progress.

As the Americans were unaccustomed to rigid airships, a party of eight experienced airmen was sent to New York to ensure there was a competent ground crew ready to receive *R34*. A supply of hydrogen to top up its gasbags was also sent.

On 1 July the airship was gassed to its limit and loaded to its full capacity. The official departure time was set for 02:00 hours the following morning, when conditions were optimum for maximum lift from the gasbags. The ship was carefully eased out of its shed by the handling party. The weather forecast was favourable and Scott decided not to wait any longer. At 01:42 hours the signal to release was given and *R34* lifted slowly into the misty night sky.

The engines were started but the craft was so heavy with fuel that even under maximum power height was gained very slowly. By the time *R34* had edged along the Firth of Forth to Rosyth and Glasgow it had struggled up to 1,300ft (396m). By daybreak the craft was following the Clyde; at 04:30 hours it was over Co. Antrim.

The crewmembers were divided into two watches. They had been issued with heavy duty flying suits, which had been redesigned to include parachute harnesses and integral life-saving collars. Life on board *R34* quickly settled down to a routine of watches, meal and rest times. The sound of jazz music could be heard wafting through the craft from the gramophone, which had been added to entertain the crew.

Over the ocean the morning fog lifted. Now it was possible to see that the airship was hanging between two cloud layers, the upper one obscuring the sun. The wireless operators were discovering that these weather conditions were causing electrostatic shocks from the equipment. But the clouds soon parted and the sun broke through. Scott was wary of the effect of superheating on the gasbags and was determined to avoid valving off any hydrogen at this early stage of the flight. Accordingly, he ordered the airship to descend back into the fog to protect it from the heat of the sun and allow the gasbags to cool down.

The routine aboard the airship was disturbed when it was just six hours out. A stowaway was discovered. W.W. Ballantyne, aged twenty, had been one of the crew members who had been left behind to save weight. Ballantyne had secreted himself on board two hours before the craft's departure, and had found a hiding place between the framework and the gasbags. He had intended to remain in hiding throughout the voyage, but was forced to reveal himself when he became ill and was hauled up before the captain.

Flight reported that aviation's first-ever stowaway had been feverish and was given medical attention, but that it had taken Ballantyne two days to recover. 'He was then made to work his way across,' the journal noted. 'He will not be allowed to make the return journey on board *R34* but will be sent home by steamer. Although as a matter of form he will have to be court-martialled, it is not thought that his punishment will be very severe.'

Ballantyne's illness was thought to be the result of his close confinement with the gas bags and he was treated by the second officer, Lieutenant H.F. Luck. After his recovery, the stowaway was put to work as a cook and was also required to operate the fuel pumps. But it soon transpired that Ballantyne was not the only stowaway. He had come aboard with the crew's mascot, a small tabby kitten called Whoopsie. The oldest airman on board, forty-two-year-old Leading Aircraftsman George Graham, accepted responsibility for looking after the cat for the rest of the voyage. Afterwards he received an offer of $1,000 for the animal, but refused to part with it.

Meanwhile, the weather was slowly deteriorating. The wind speed was rising as a storm approached and all the airship's five engines were eased up to full power. By the next morning *R34* was halfway across the ocean. Although the weather continued to worsen, there were some breaks in the murk, enabling the airship's crew to get a view of the sea below.

Towards the end of the day the south-easterly wind had increased speed to about 50mph (80kph), causing the airship to crab its way forwards. Scott's response was to try to avoid the worst of the wind by climbing, but this proved impossible. However, the cloud had largely moved away. The clear sky enabled the crew to see that the sea below was spotted with icebergs, one of them estimated to be 150ft (46m) high.

As *R34* neared Newfoundland the fog returned to envelop the airship completely. Inside the control cabin there were other concerns. The ship carried no fuel gauges, but the dipsticks in the tanks showed that only 2,200gal (10,000lit) of petrol remained. With further strong headwinds expected along the coast it was looking less and less likely that the airship would be able to reach New York as planned.

Land was sighted at 12:50 hours and the ship was following the coast southwards. Fuel was a mere 500gal (2,270lit) and Scott brought the airship down to 800ft (244m) to avoid the worst of the wind. From this height *R34*'s crew had a clear view of North America as it slowly unwound beneath them. This brought some relief as the airship had been airborne for four days and the crew was beginning to tire.

Meanwhile, preparations were being made in Boston for an emergency landing there, but the airship continued. Each fuel tank was carefully inspected and whatever remained was collected and poured into the main tanks to keep the engines running. It was then that Scott made the decision to continue to the agreed alternate landing site at Mineola, Long Island, New York. In the last hour of the flight, the crew made themselves presentable ready to be acclaimed as the first aviators to make the Atlantic crossing from east to west.

Even though Mineola was not the craft's scheduled destination, there was a band on hand to play *God Save the King* together with thousands of spectators. Grandstands had been erected around the landing ground and car parks were provided. At 0900 hours the citizens of Garden City and Mineola had left their houses to watch the airship circling overhead as it prepared to land. 'The number was clearly visible on its side and her great bulk was gleaming in the morning sunshine,' *The Times* reported.

A few minutes later, according to the paper, 'a shout went up as a tiny object detached itself from the rear gondola and floated earthwards'. This was Major John Pritchard. He had strapped on a parachute and, as the airship circled the landing area, dropped to the ground to give landing instructions. This made him the first man to arrive in America by air. He hastily arranged the ground crews and helped ease the airship down. 'He came to earth in front of the grand stand and was taken to headquarters on a motor cycle,' *The Times* reported.

R34 landed at 09:54 hours after a flight of 108 hours 12 minutes. There were 140gal (636lit) of fuel left in the tanks, sufficient only for another two hours' flying at reduced power. The airship was to remain in America for just three days. During this time the crew could rest and enjoy hot showers. The people of New York lavished generous hospitality on the crew and they were bombarded with offers of invitations to formal functions during their visit to mark their historic achievement.

R34 was moored by a three-wire system at the bow, the lift from the gas bags keeping the mooring wires taut. The airship's engineers stayed with the craft to overhaul the engines and prepare them for the return voyage. They found that that no repairs were necessary and that the engines had performed well. The propellers had, however, accumulated a thick coating of engine oil and this was removed free of charge by a local company.

The crew returned to the ship and provisions were loaded for the return flight. The final preparation was to replenish the gas bags. This was done by using thousands of cylinders of hydrogen gas. As with the outward flight, *R34* would be gassed to capacity and its departure would take place during the coolest part of the day. Accordingly, the craft departed on its return flight at 23:54 hours on Wednesday, 10 July. The huge crowd that had greeted the airship on its arrival in the USA was still there and gave a huge cheer as *R34* rose serenely into the night sky.

The wind had picked up before the launch and was now gusting at 30mph (48kph). Despite the concern this caused the airship's crew, *R34* cleared the landing field and made its way eastwards. As a gesture of gratitude towards the city, *R34* headed for the bright lights of New York.

General Maitland was to write later:

> New York at midnight looks wonderful from above. Miles and miles of tiny bright twinkly lights – a veritable fairyland. The searchlights at first make a very unsuccessful search for us but finally get us fair and square. We are over Fifth Avenue. Times Square and Broadway present a remarkable sight. We distinctly see thousands of upturned faces in spite of the early hour, one o'clock in the morning, and the whole scene is lit by the gigantic electrical signs which seem to concentrate about this point.

The ship made its way up to an altitude of 2,000ft (610m) as Scott was unsure of the height of the city's skyscrapers. Despite the late hour, thousands of well-wishers took to the city streets and rooftops to wave at the airship as it passed overhead. The craft then turned out to sea and headed for home.

R34 made good progress during the night as it now had the wind behind it. Speed increased to 90mph (144kph), even though the forward engine was shut down. The crew was unprepared for the swiftness of the eastward journey and it seemed that fuel was being burned at a much slower rate than during the outbound voyage. The return home was uneventful, and the standard ship routine continued.

Maitland kept a detailed log of the voyage and this was released to the press soon after *R34*'s arrival at Pulham in Norfolk. This describes in some detail the ever-changing weather and unusual cloud formations observed by the general. But it also underlines some of the big differences between the conditions faced by Alcock and Brown compared with the comparative luxuries enjoyed by Major Scott and his crew.

Instead of hurried snacks of sandwiches and chocolate, *R34*'s officers were able to sit down to rather more relaxed meals. On the first day of the return flight they enjoyed a lunch comprising cold Bologna sausage and pickles with stewed pineapple washed down with a 'much-appreciated ration of rum'. 'Mealtimes,' Maitland noted, 'are always most welcome and give the more responsible members of the crew a much-needed interval.'

In-flight music was another luxury enjoyed by *R34*'s crew. Maitland wrote: 'We are well-equipped with little luxuries on this return voyage having learned a thing or two about what is necessary and what is not.' Accordingly, the craft carried a better gramophone than it had on the outward journey and the crew seems to have made good use of it. On one occasion, while descending the ladder to the fore car one lunchtime, Maitland was amused to observe the second officer and the meteorological officer 'doing quite a nice one-step together'.

The only problem occurred when an engineer fell against the clutch of an engine, causing it to race to its destruction when a connecting rod broke and punched through the side of the cylinder block. In-flight repair was not possible, so the engine was stopped. This did not slow the airship, but it was decided to abandon the idea of over-flying London and head directly for home. The Air Ministry, however vetoed this plan due to deteriorating weather at East Fortune and ordered Scott to head for Pulham.

A few hours later a message from East Fortune confirmed the weather had improved there. A request was put to the Air Ministry to allow the ship to return to East Fortune, but this was turned down. No reason was ever given for this change in plan and no explanation was ever found. The airship wafted over the English countryside and came rather quietly to Pulham Air Station at 6.57 hours to be welcomed by the RAF personnel. It was a lower-key reception than the ship had received at New York and might have been expected at East Fortune. The return journey had taken three days, three hours and three minutes. The ship had travelled some 7,420 miles (11,870km) on this voyage at an average speed of 43mph (69kph).

The Times' correspondent was one of many people at the airship station at Pulham St Mary near Diss awaiting the *R34*'s arrival. 'She might have been a cloud herself in shape and hue and seemingly slow movement,' he reported. 'She was creeping out of the north-west, flying at 1,500ft and not until she was almost overhead could a sound be heard from her.'

On the ground, the mooring parties had assembled and took up their position as the great airship manoeuvred above them. Ropes were thrown from the craft and cable ends joined to enable the mooring party to take control. Then, according to *The Times* reporter, 'the sudden ejection of her water ballast and the emptying of her water storage tanks occurred just as the RAF band, much depleted by week-end leave, struck up *See the Conquering Hero Comes*'. The paper reported that the airship had returned with 1,000gal (4,546lit) of fuel in its tanks, but with one of its five engines not running.

Once the craft had been guided into its shed, General Maitland leaned out of the forward gondola and handed *The Times*' representative a letter addressed to his editor. Maitland, wearing RAF uniform, was the first to disembark and was followed by Major Scott. They were met by the station commander, Lieutenant Colonel Boothby, after which they and the rest of the crew sat down to breakfast.

During *R34*'s voyage a progress board had been erected outside Marconi House in London's Strand to give passers-by information about the airship's position from the messages received from it. There were messages of congratulation from Winston Churchill and senior air force officers.

There had been thirty officers and men aboard *R34* on its outward journey and thirty-one on the return voyage. Lieutenant Commander Zachary Landsdowne of the US Navy had been aboard on the trip to New York and on the return voyage his place was taken by Major W.H. Hemsely of the US Army Air Service. Scott, Maitland and Lieutenant John Shotter, *R34*'s engineer officer, travelled from Norfolk to London to represent the airship's crew and all three gave press interviews.

Flight thought Shotter, twenty-nine, to have been one of the hardest working crew members. 'He was responsible for the care of the five Sunbeams,' the journal noted, and, in fact, for everything of an engineering nature on the airship.' It was revealed that Shotter's wife had been taken ill on the eve of the flight, but he had elected to remain on board. 'Under such mental pressure his devotion to duty deserves the highest praise,' *Flight* declared.

Scott told the *Daily Telegraph*'s H.C. Bailey that the voyage had been an uneventful one. On the return journey the airship had reached speeds of up to 80mph and, despite the fog that prevented navigational sightings being taken, the directional wireless had enabled the airship to remain on course 'without difficulty'.

Scott was quoted as saying that *R34* had shown that the large airship 'is the type of aircraft for deep sea work'. 'Before long,' he forecast, 'I hope that we shall have airships of a size and speed which will enable them to sustain a rate of 70–80 mph.'

Maitland told the *Telegraph* that one of the objects of the flight had been 'to demonstrate what airships could do in long-distance flying oversea with an ultimate view to their commercial use'. Airships, he predicted 'will undoubtedly be employed for commercial purposes for very long journeys over sea and land. They will not conflict with the aeroplane and seaplane. Airships will make long voyages and from their termini aeroplanes will radiate on short, quick journeys'. *Flight* commented that *R34* had 'erected another milestone ... on the long road of aerial navigation'.

The enthusiastic press coverage contrasted with the lukewarm official attitude towards the airship and its crew. *Flight* reported that they were met at Liverpool Street station by Mrs Churchill and representatives of the Air Ministry and the Air Staff. The reason for this low-key welcome would become clear within the next two years. In May 1921 the government cancelled all airship development for financial reasons. Military airships were scrapped and *R34* was eventually broken up after being damaged in a crash on the Yorkshire moors.

Edwin Colston-Shepherd, author and air correspondent of *The Times*, talked about bureaucratic apathy towards *R34* and its crew. They had, he wrote, taken all the chances that went with a first attempt and had made the dangers look small and ordinary 'by their diligence and attention to detail'. The accidents that would befall later airships showed how great was their achievement. 'Nothing,' Colston-Shepherd declared, 'can rob the feat of its own inherent greatness and none can dispute the courage of men who did so fine a job so greatly and well.'

It would be another decade before a British airship again attempted to cross the Atlantic. Meanwhile, the airship station at Pulham St Mary was put on a care and maintenance basis and later closed. What the locals called the flying pigs had been grounded for good. Scott and Maitland continued their advocacy of airships, for which both were to pay with their lives. Scott was killed when the *R101* was wrecked near Beauvais in 1930, while Boer War veteran Air Commander Maitland died in 1921 when *R38*, *R34*'s sister ship, suffered structural failure and broke up in mid-air over the Humber.

Tiny – The Inside Story

The R33 class airships originally designed for the Royal Naval Air Service were big. At 643ft (196m), *R34*'s envelope was as long as a battleship or two football pitches laid end to end. The diameter was 79ft (24m) and several Vimys would have fitted comfortably within the envelope. Inevitably, the *R34*'s crew would nickname the monster 'Tiny'.

Like *R33*, *R34* was semi-streamlined fore and aft, with the middle section being straight-sided. The metal framework was varnished to prevent atmospheric corrosion and heavily braced by wire. Strips of linen were stretched between each pair of frames and attached by laces. Narrow strips were then glued over the lacing, and the envelope that covered the structure was treated with dope containing aluminium powder. This was intended to reflect sunlight and so reduce the risk of superheating the gas inside.

The nineteen gasbags were located within the chambers formed by the metal structure. They comprised a single layer of rubber-proofed cotton cloth, varnished and lined with goldbeaters' skin. Each gasbag was contoured to fill the available space and surrounded by cord mesh to prevent chafing against the girders. Suspended beneath the main body of the airship by long wooden struts and braced by wires were four small gondolas. The control car was well forward and contained the steering and elevator wheels, the gas-valve controls, the engine telegraph, navigational instruments and the toggles controlling the emergency forward water ballast.

A ladder, protected from the elements by a streamlined canvas cover, connected the control-cabin with the keel. A similar cover enclosed the control-cable connections leading into the hull. The aft section contained an engine in a separate structure to stop vibrations affecting the sensitive radio direction finding and communications equipment. The small gap provided to isolate the main structure from engine vibrations was faired over so that the 50ft (15m) gondola appeared to be a single structure.

The craft was powered by five 275hp (205kW) twelve-cylinder, water-cooled Sunbeam Maori engines that had been designed specifically for airship use. The superior Rolls-Royce units that had powered earlier rigid airships were now reserved for aeroplanes. The Maori units were designed to run at 2,100rpm, but rarely exceeded 1,600. In *R34* there was one in the aft section of the control car and two more in a pair of power cars amidships. These engines each drove a pusher propeller via a reversing gearbox that facilitated manoeuvring during the mooring process. Two more engines were mounted in a centrally located aft car and were geared together to drive a single pusher propeller.

If an engine stopped during flight, the propeller was disconnected to allow it to rotate freely in the airstream and reduce resistance. If the craft was required to remain stationary, as for landing, a special brake was provided. If the airship was still moving forwards, the engine could be started by releasing the brake, re-engaging the clutch and allowing the airstream to turn the propeller.

There was a lot of space within the envelope. Running almost the entire length of the ship was a long corridor comprising a succession of A-shaped frames standing on the two lowest girders. Three smaller ones effectively fenced off the surrounding gasbags. At its widest part, this corridor was about 10ft (3m) across, narrowing towards the extremities. Leading to the wing and after cars were narrow ladders, fully exposed to the force of the elements.

Tests with *R33* showed that the two craft could turn within 4,100ft (1,250m). The effect of the slipstream from the after propeller acting on the rudder was so strong that, with the forward engine stopped and the wing propellers both running in reverse, it was possible for the airships to pivot virtually on the spot.

R34 Specification

Overall length	643ft (197.8m)
Envelope diameter	79ft (24.3m)
Maximum speed	62mph (99kph)
Power plant	five Sunbeam Maori twelve-cylinder, water-cooled engines, each developing 275hp (205kW)
Gas capacity	1,950,000 cu ft (55,218 cu m)

Chapter Seven

Wings of the Morning

In May 1919 a French-born New York hotelier offered a prize of $25,000 to the first aviator to fly between New York and Paris. It was to be nearly a decade before a serious attempt was made to claim the cash, but the result was to make a whole nation aware of the potential of commercial aviation virtually overnight.

Raymond Orteig had emigrated from France to the USA when he was twelve years old. He arrived in New York with just 13 Francs in his pocket, but at least he had an uncle living in the city.

The boy got a job as a porter in a restaurant and worked so diligently and so hard for the next twenty years that he was able to buy the hotel in which he was then employed as *maître d'*. During the First World War the Lafayette Hotel became a popular haunt for airmen, particularly French officers on temporary duty in the US. This gave Orteig an interest in aviation and in the idea of the USA and the country of his birth being linked by air.

Orteig offered his prize via the Aero Club of America at a time when the Atlantic had yet to be flown in a single hop. He set a time limit of five years for this feat to be achieved but aeronautical technology had yet to reach the point where a realistic response to Orteig's challenge could be mounted. The prize remained unclaimed but in June 1925 Orteig repeated his offer. He even deposited the $25,000 in a local bank under the supervision of a seven-member board of trustees.

By September 1926 interest in the prize was mounting. One unsuccessful attempt had already been made when a lanky twenty-five-year-old air mail pilot from Missouri decided he should have a crack at it.

So it was that, little more than six months later, according to legend, Charles Lindbergh squinted up at the sky above Roosevelt Field, Long Island, and drawled: 'Well, I guess I might as well go,' and thirty-three hours and thirty minutes later arrived in Paris.

Over the next forty-five years it was sometimes hard to separate legend from fact. But it was indisputable that the man called 'the Flying Fool' by the news media had become the first to fly the Atlantic solo, the first to fly from one continental land mass to another and the first to fly from New York to the French capital.

Like Alcock and Brown eight years before, Lindbergh received a hero's welcome, he won a valuable cash prize and his success was largely due to his professionalism and attention to detail. But even Alcock and Brown would have been hard-pressed to imagine the consequences of Lindbergh's flight.

In the immediate aftermath he was presented to two monarchs, one of them King George V. Two presidents, those of France and the USA, wanted to shake his hand. When Lindbergh visited London the crowds waiting to welcome him at Croydon airport were so enthusiastic that he had to take-off again immediately after touching down to avoid running into them. On his return home, the ticker tape parade through the streets of New York was unprecedented, with estimates of the crowd lining the route varying between three and four million people. It cost the city $16,000 to clear up afterwards.

Perhaps Lindbergh, rather like Raymond Orteig, had become the living embodiment of the American dream – the young man who worked hard and by sheer guts and determination had achieved fame and fortune. But the longer-term consequences for commercial aviation were certainly profound. Lindbergh's achievement had caught the public's imagination to the extent that, virtually overnight, America was transformed from an aeronautically ambivalent nation to one that had suddenly come alive to the possibilities offered by aviation.

It was as if Alcock and Brown had never been. And never mind the US Navy's NC–4, the British and German airships, *R34* and *LZ-126*, or, indeed, the US Army's World Cruise of 1924 that, admittedly made the trans-Atlantic crossing in convenient hops. The fact was that, even though around 120 men had crossed the Atlantic before Lindbergh, most Americans thought him to have been the first.

In 1927 most of Lindbergh's fellow citizens used the nation's extensive rail network to travel the USA; airplanes were there to carry the mail and perform tricks. But such was the energy and enthusiasm released by the Lindbergh flight that the US aviation industry set off at a furious pace to catch up. Within six years Boeing had produced what is regarded as the first modern airliner, the Model 247, to be followed shortly afterwards by the Douglas DC-2. The follow-up DC-3 revolutionised commercial aviation and helped win the Second World War.

Charles Augustus Lindbergh had been born in Detroit in 1902. His father, also called Charles, was of Swedish ancestry and was a politician, having been elected to the US Congress for the Sixth District of Minnesota where the family lived. Later, Charles Lindbergh Senior failed to gain election to the Senate and the state governorship. It may be that his antagonism to US intervention in the First World War went against him, but isolationism was a cause Charles Junior would espouse two decades later.

Meanwhile, the young man grew up lean and tall, and mechanically minded. He drove his father's Model T Ford during his 1914 election campaign. After an early interest in a farming career, the younger Charles decided he wanted to drop out of college and learn to fly instead. Later, he would write that when he told his mother, chemistry teacher Evangeline, who was Charles Senior's second wife, she was resigned to it. 'All right,' she had said, 'if you really want to fly that's what you should do.' Later, she would fly with her son barnstorming in southern Minnesota and in his mailplane, riding on the sacks in the DH.4 between Chicago and St Louis.

In 1922 young Lindbergh learned to fly at a school in Lincoln, Nebraska, and joined the ranks of the airmen who barnstormed their way around the countryside entertaining crowds of spectators with their daredevil antics. He soon had his own aircraft, a Curtiss JN-4 Jenny. It cost him $900 but he soon conceived the idea of being paid for his flying and in 1924 he enrolled in the Army for a one-year stint. Although he subsequently returned to barnstorming, he remained in the Army Reserve and by the time of his flight to Paris had attained the rank of captain.

In October 1925 the tall, fair-haired Lindbergh, now universally known as 'Slim', met war veterans Bill and Frank Robertson, who had established a business offering flying tuition and charter flights. Earlier that year Congress had passed an Act making it possible for private operators to fly the mail on individual routes that had previously been operated by the Post Office. The Robertsons made a successful bid for the Chicago–St Louis route, which they operated with intermediate stops at Peoria and Springfield using a Liberty-powered DH.4B.

Flying in all weathers and often at night in unreliable aircraft equipped with inaccurate and meagre instruments was risky work. By autumn 1926, Lindbergh had not only become chief pilot of the Robertson Air Corporation, but had also made two escapes from crippled aircraft by parachute to add to the two he had made during his Army service.

He had also made the decision that would alter his life.

In his account of the flight to Paris, published twenty-six years later after a decade of writing and rewriting, Lindbergh described how, one night in September 1926 as he was flying to Chicago, he amused himself by wondering what it would take to get to Europe. How much fuel? How long could the engine run continuously? What type of aircraft would be best?

He convinced himself that with the right machine he could really demonstrate the potential of aviation. Somehow, he would have to make the businessmen who would be his potential sponsors understand the possibilities of flight. But he kept coming back to one central question: why should he not fly from New York to Paris.

'I'm almost twenty-five,' he told himself, 'I have more than four years of aviation behind me and close to 2,000 hours in the air. I've barnstormed over half the forty-eight states. I've flown the mail though the worst of nights.' And during his time in the Army he had learned the basic elements of navigation. 'I'm a captain of the 110th Observation Squadron of Missouri's National Guard,' he told himself. 'Why am I not qualified for such a flight?'

Even at this early stage Lindbergh was convinced that a single-engine aircraft flown solo was more likely to get the job done than a fuel-hungry, multi-engine machine with several crew members. The fate of Rene Fonck and his crew (*see* Chapter 8) had reinforced this belief. He had been impressed by claims made for the Bellanca monoplane and he was certain that the best power plant for the job was the 220hp nine-cylinder Wright Cyclone radial.

But, as he put it, 'above all else looms the question of finance'. He reckoned he could probably find a couple of thousand dollars, but the bulk would need to come from corporate sponsors, local St Louis businessmen. The way he figured it, a non-stop flight between America and Europe would demonstrate the potential of air transport and help place St Louis in the forefront of aviation. Not only that, but a successful flight would more than cover its costs thanks to the Orteig prize.

But potential sponsors were sceptical that the New York–Paris flight could be accomplished by a single-engine aircraft. And when he approached some of the bigger airframe manufacturers the response was the same. Fokker flatly refused to build a single-engine aircraft for the flight and the Wright-Patterson company took a similar view. But such was the persistence of the young air mail pilot that he was able to find the backers. Harold Bixby, St Louis bank president and head of the local chamber of commerce, was particularly impressed by Lindbergh's proposition. His support was to prove crucial.

With the finance as good as settled, Lindbergh was now free to speak frankly with the aircraft manufacturers. But the new owners of the Bellanca company, as well as another respected manufacturer, Travel Air, were not interested in selling an aircraft to be piloted by Lindbergh. But then he heard of a new high-wing type being used on one of the mail routes and which had attracted favourable comment.

In February 1926, therefore, Lindbergh visited the Ryan factory in San Diego. Quickly he and its management got down to cases: Ryan wanted $6,000 for the aircraft less engine or $10,580 with the latest J-5 Whirlwind engine fitted. Lindbergh was able to wire his backers in St Louis with the details. 'Recommend closing deal,' he concluded. The reply came the next day: 'Close the deal.'

'The chafing, frustrating weeks of hunting, first the finance and then for a 'plane, were over,' he wrote. 'I can turn my attention to the flight itself – to the design and construction of the 'plane, to outfitting it with instruments and emergency equipment, to studying navigation and the weather conditions I'm likely to encounter along the route.'

The construction of the aircraft, now officially named *Spirit of St Louis* at the suggestion of Harold Bixby, moved along rapidly. 'Everyone is taking a personal interest in the flight,' Lindbergh noted. 'Hours of overtime labour have become normal and voluntary.' Ryan's chief engineer, Donald Hall, often started work at 05:00 hours to inspect the previous day's progress before the rest of the workforce arrived. Lindbergh, meanwhile, was starting to think about the flight itself and making preparations. The aircraft would carry fuel enough for 4,100 miles 6,560km), giving a reserve of 500 miles (800km).

There was great excitement when the Whirlwind engine arrived at the factory. Lindbergh would remember the moment vividly:

> We gather around the wooden crate as though some statue were to be unveiled. It's like a huge jewel, lying there set in its wrappings. Here is the ultimate in lightness of weight and power – 223 horses compressed into nine delicate fin-covered cylinders of aluminium and steel. On this intricate perfection, I'm to trust my life across the Atlantic Ocean.

Flight testing was largely uneventful. First, the aircraft with its increased 46ft (42.5m) wingspan, had to be eased out of the factory on the first stage of its journey to Ryan's airfield at Dutch Flats. It was exactly sixty days since the formalities had been completed for the purchase of the aircraft and work on it begun. Lindbergh was entranced by its appearance. 'What a beautiful machine it is, resting there on the field in front of the hangar, trim and slender, gleaming in its silver coat!' he enthused. 'All our ideas, all our calculations, all our hopes lie there before me, waiting to undergo the acid test of flight. For me, it seems to contain the whole future of aviation.'

Lindbergh was nothing if not pragmatic and refused to have anything on board the aircraft that was not strictly necessary for the flight to Paris. That included a windscreen. There would be no forward view because the high wing was set so close to the fuselage to reduce drag that there was no room for a windscreen. There were two rectangular windows either side of the pilot's seat through which he could poke his head if he really needed to look ahead – he was planning on only one landing, after all – and there was also a periscope.

As it happened, one of the Ryan mechanics was a young man called Douglas Corrigan, who will reappear later in this story. Lindbergh acted as his own test pilot and over the next few days made a series of test flights. Lightly loaded, the aircraft seemed very sprightly. He established that it was capable of 128mph (205kph) and found no major faults.

On 10 May *Spirit* left San Diego for the 1,500-mile (2,400km) flight over the Rocky Mountains to St Louis. By 0800 hours on the 12th the aircraft was over the city so that Lindbergh could show it off to his backers in the business district. But there was little time for celebration or even rest. Within fifteen minutes of landing he was off again, this time bound for Roosevelt Field, New York, the jumping-off point for the flight to Paris.

Lindbergh estimated that several thousand people had gathered to welcome him, including reporters and cameramen. Later there was a press conference and Lindbergh – now dubbed the 'Flying Fool' by reporters – realised that news media interest was something he would have to get used to. He would also have to get used to waiting for favourable weather. It was not until 20 May that the weather looked like being good enough for the flight.

Yet the evening before there had been thick coastal fog over Nova Scotia and Newfoundland and Lindbergh had planned to go to the theatre with friends. But a late report of good weather over most of the North Atlantic and a promise of the fog lifting persuaded him that, if he were to forestall his rivals, it was now or never. Instead of going to the theatre, he decided to spend the time on last-minute preparation. Morning brought low clouds and rain and the ground was soft and muddy, not ideal for the start of a trans-Atlantic flight. But Lindbergh ordered the *Spirit of Louis* to be rolled out of its hangar and its fuel tanks topped up.

It was still dark when the aircraft, its engine protected by a tarpaulin wrapped around it, was towed out to Roosevelt Field. Escorted by policemen on motor cycles, reporters, fellow aviators and onlookers, the *Spirit* was dragged tail-first by a lorry. It did not look like an auspicious occasion. 'It's more like a funeral procession than the beginning of a trip for Paris,' Lindbergh observed. Indeed, the enterprise appeared doomed from the time the aircraft, its silver bullet-shaped fuselage dripping with rain, eased forwards through the puddles of water collecting on the runway and towards the telephone wires at the end, which looked uncomfortably close.

The speed built up as Lindbergh struggled to keep the machine straight. The wheels left the ground, touched again and the aircraft shuddered. But flying speed was gained little by little. The *Spirit* was airborne, although the telephone wires were cleared with just 20ft (6m) to spare. From then until he was over the ocean Lindbergh continually studied the ground below seeking suitable landing places in case his engine failed. Just fifteen minutes into his journey, Lindbergh became aware that he was not alone in the sky. Another aircraft was flying alongside the *Spirit* with cameras poking through its windows. It had not occurred to him that the newspapers would hire aircraft to follow him on his way to Paris. He felt slightly irked, but his irritation did not last: once they had turned back the sky belonged to him.

By the end of the second hour the *Spirit* was over New England, having travelled 100 miles (160km) with, as Lindbergh told himself, 3,500 miles (5,600km) to go. By the third hour the aircraft was flying over the sea and out of sight of land on its way to Newfoundland. He dropped down until he and the *Spirit* were skimming 20ft above the waves. But by the fourth hour Lindbergh was beginning to feel tired. He started thinking of sleep, something he should not be doing and certainly not with just one-tenth of his journey completed.

Nevertheless, the desire to sleep would be his constant companion from then on, its insidious presence in his mind growing more and more insistent as the hours without sleep piled up. For now, though, he banished the thought by focussing on other needs as he sipped water from his canteen. Beneath the canteen were stored his rations for the journey: five sandwiches in a brown paper bag. But, Lindbergh reasoned, it would probably be wiser not to eat as it was easier to stay awake on an empty stomach.

In the fifth hour he forgot about feeling tired. He was over land again: Nova Scotia. The *Spirit* was flying at 1,000ft and the first test of Lindbergh's navigational skills had been successfully passed: he was just 2 degrees out. It was less than half what he had allowed for. And by 1300 hours he became exhilarated by the thought that he had left New York at breakfast time and had reached Nova Scotia in time for lunch. But he was still not hungry.

He was, though, aware of the draught blowing through his cockpit and ruffling the charts. He decided not to close the side windows; the saving in drag would not be worth the loss of ventilation. Now he was flying into a rain squall and that meant a small detour. The squalls eased going into the seventh hour, when the patches of blue sky grew larger.

By the ninth hour *Spirit* was over the sea again: the next land would be Newfoundland. The weather improved. The desire for sleep returned. If only he could throw himself down on a bed! Later, Lindbergh would recall: 'My eyes feel dry and hard as stones. The lids pull down with pounds of weight against their muscles. Keeping them open is like holding arms outstretched without support.' And night was falling. Lindbergh took the *Spirit* up to 300ft (90m) over the waves and tried to push aside the feeling of sleepiness. He shook his head, stamped his feet on the floor and breathed deeply.

Down below the surface of the sea was dotted with stark white icebergs. Lindbergh's thoughts turned to what he would do if his engine failed now. As he neared the Avalon Peninsula, jumping off point for Alcock and Brown's trans-Atlantic flight eight years earlier, Lindbergh's flight entered its twelfth hour. The wind was blowing him along at, he reckoned, the rate of a mile every thirty seconds. Soon he was over St John's and heading for the sea again. Now, he told himself, he had reached the point 'where real navigation must begin'. And he believed he was 90 miles (144km) south of the great circle route he had planned. The next land he would see would be Ireland.

For now, at least, he was on schedule, there was plenty of fuel in the tanks, the engine sounded stronger than ever and he had a tailwind. But he was facing the monotony of the long oceanic flight and the ever-present desire for sleep. The clouds were building up like a mountain range and it was beginning to get hazy. Soon the haze was replaced by storm clouds far taller than *Spirit* could reach, even though Lindbergh's altimeter was reading 10,500ft (16,800km).

As Lindbergh tried to plough through the clouds, ice started forming on the aircraft's wings and propeller. He had to turn around, get back into clear air – quickly before his instruments became unreliable due to the accretion of ice. Within ten minutes he was clear of the thunderclouds, but he now had to decide the best way to fly around the storm and resume his previous course. He had to go south.

With the approach of dawn the moon had become visible and the clouds were thinning, although more were beginning to form ahead. With the passage of another hour, the fifteenth since his departure, Lindbergh was reflecting that *Spirit of St Louis* was nearer to Ireland than to New York. But it still had to get back on course.

By the seventeenth hour the ice had almost gone and the temperature seemed to have risen. Lindbergh could discard his mittens and unzip his flying suit. But the desire for sleep returned stronger than ever. 'I've lost command of my eyelids,' he wrote later. 'When they start to close I can't restrain them. They shut and I shake myself and lift them with my fingers.'

When he reached the halfway point of his journey Lindbergh wondered if he should celebrate by eating a sandwich. He decided he was not hungry or even thirsty. There were still eighteen hours to go. With the sun rising the *Spirit of St Louis*' cockpit began to warm up until it became uncomfortably hot. By midday the desire for sleep was stronger than ever and Lindbergh had to fight it off again. But sighting schools of porpoises and then birds told him that he was nearing land.

During the twenty-seventh hour he sighted several small fishing boats. There was no telling where they had come from – Ireland, England, Scotland or France – but at least they represented some form of human contact after hours of solitude. As he flew over one of the boats a head appeared at a cabin porthole staring up at him.

Lindbergh glided down to within 50ft of the boat, throttled back his engine and shouted: 'Which way is Ireland?' He passed over the boat three times but there was no response. *Spirit of St Louis* resumed its original course, but an hour later Lindbergh had his answer. He had sighted land: it had to be Ireland. It was in fact, the south-west coast between Cape Valentia and Dingle Bay, almost exactly on the route he had planned earlier. He was on course and two hours ahead of schedule!

By the thirtieth hour the Cornish cliffs became visible, rising sheer out of the sea. Soon he was flying over the English Channel south of Plymouth with the coast of France just visible. As the sun was setting *Spirit of St Louis* was over Cherbourg. Another hour and the lights of Paris should be visible and Lindbergh began reflecting on the kind of reception he could expect when he landed.

But the reality was very different. Having recognised the airfield of Le Bourget and flown over the Eiffel Tower he landed – 'not a bad landing,' he thought – and came to rest. He started to taxi towards the hangars but then noticed that 'the entire field ahead is covered with running figures!' He had barely shut the engine down when his open cockpit windows were blocked with faces and his name was being shouted over and again. The surge of people around his aircraft tearing at the fabric for souvenirs was so great that Lindbergh feared for its safety.

He opened the door but was lifted up by the crowd pressing around the aircraft. For half an hour he was carried shoulder high. Lindbergh was their hero. They had been following his progress and, anticipating his arrival, the French newspapers had rushed out special editions. Rescued by French air force officers, Lindbergh spent the night in the US Embassy. By morning the building was besieged by well-wishers calling for him to come out on to the balcony. Lindbergh was a celebrity and the people of Paris were the first to take the lanky young man with the shy grin to their hearts.

After a week of public appearances, he and *Spirit of St Louis* were in the air again, heading this time for Brussels and London for more tributes. 'It might almost be said,' mused *The Times*' air correspondent, Edwin Colston Shepherd, 'that it was Europe that discovered the quality of Lindbergh. His own country awoke slowly to the significance of his flight. The tributes of France aroused the United States. Equal warmth in Brussels and then in London showed the United States that the persistence of one of her unknown sons had presented her with a new hero.'

The nation seemed to be fully aware of this by the time Lindbergh returned. The US Navy sent the cruiser *Memphis* to bring him home. The RAF dismantled the *Spirit of St Louis* and crated it for shipping. He was given a hero's welcome in Washington, where President Coolidge greeted him with an address of welcome. Between July and October Lindbergh took *Spirit of St Louis* on a 22,000-mile (35,200km) tour of all the forty-eight states. Then, in December, he left on a further tour, of Central America and the Caribbean.

Spirit then went into well-earned retirement at the Smithsonian Institute. But there was no retirement for Lindbergh for his achievement seemed to have started something. As airline historian Ron Davies observed, it:

> … probably produced greater mass enthusiasm than any other event in the history of air transport – or, indeed, the history of aeronautics. It gave the struggling young industry the extra stimulant which it so badly needed. The immediate effect was not a direct one, only that the American people were suddenly seized with the idea that the aeroplane was a safe, speedy and useful vehicle. This realisation awoke the interest of ordinary folk who just wanted to experience the thrill of flying and more important, perhaps, the awareness of businessmen that aircraft building and operating could be a profitable investment. There had been nothing like it since the railway boom of the previous century.

Lindbergh threw himself into the development of commercial aviation for Transcontinental Air Transport (later TWA) and later Pan American, which was run by his friend, Juan Trippe. He undertook many route-proving flights and during the Second World War helped test service aircraft under operational conditions.

Charles Lindbergh died in 1972. On his tombstone, by his own instruction, there appeared the following inscription taken from the 139th Psalm:

> If I take the wings of the morning and dwell in the uttermost parts of the seas …

According to Ron Davies, he had been, arguably the greatest pilot of all time who 'made a priceless contribution to the advancement of aeronautics'.

Ryan NYP

When he saw the completed Ryan NYP (New York to Paris) for the first time Charles Lindbergh thought: 'For me, it seems to contain the whole future of aviation.'

Ryan Airlines had been established in 1925 by Franklin Mahoney and Charles Ryan. The NYP was based on the company's high-wing, strut-braced Brougham monoplane, itself a development of the company's M-1 mail plane, adapted to Lindbergh's requirements by the company's chief engineer and designer, Donald A. Hall.

These modifications included a 2ft (0.6m) longer fuselage and a 10ft (3m) increase to the wingspan. The structure was modified to house the fuel required for the flight, which meant that the passenger accommodation in the forward fuselage was replaced by a large fuel tank. In fact, the NYP had two fuel tanks in the nose plus three more in the wings, giving a total capacity of 155gal (705lit). Behind the engine was a 28gal (127lit) oil tank.

Structurally the aircraft was conventional for its day, being constructed of steel tubes covered in wood and fabric. The wing featured box section spars and wooden ribs. The leading edges were covered with plywood for additional strength. The wings were strut-braced to the lower fuselage and one of the aircraft's most noticeable features was the wide-tracked undercarriage, heavily braced to enable it to bear the additional weight of fuel.

The cockpit was moved further to the rear so that the pilot sat directly behind the 209gal (950lit) fuel tank. To balance the additional length of the fuselage the Wright Whirlwind engine was moved forward for balance to permit the fuel tank to be installed at the centre of gravity.

The cockpit was spartan and lacked even a windscreen. Measuring 36in (94cm) in width, 32in (81cm) in length and 51in (130cm) in height, it was so cramped that Lindbergh could not stretch his legs. The instrument panel housed fuel pressure, oil pressure and temperature gauges, clock, altimeter, tachometer, airspeed indicator, bank and turn indicator, and a liquid magnetic compass.

The main compass was mounted behind Lindbergh and he read it by using the mirror from a women's make-up case that was fixed to the roof with chewing gum. Lindbergh also installed an earth inductor compass newly developed by the Pioneer Instrument Company. This facilitated navigation while taking account of the Earth's magnetic declination.

The aircraft was completed in late April 1927. It was painted silver with black lettering, including registration number N-X-211.

Ryan NYP Specification

Length	27ft 7in (8.4m)
Wingspan	46ft (14.2m)
Height	9ft 10in (3m)
Maximum gross take-off weight	5,250lb (2,386kg)
Power plant	nine-cylinder Wright Whirlwind J-5C air-cooled radial engine developing 223hp (166kN)
Cruising speed	10mph (169kph)
Range	4,100 miles (6,600km)

Chapter Eight

The Baron Flies West

Ehrenfried Günther Freiherr von Hünefeld might have looked like a typical Prussian Army officer and, indeed, he had been born in Konigsberg, but the monocle he wore was necessary for a man blind in his left eye and near-sighted in his right.

His childhood was characterised by serious illness and it was poor health that prevented von Hünefeld from joining the German Air Service on the outbreak of the First World War. Even so, he volunteered to serve in the Army as a motorcyclist, but his career was cut short when he was wounded in Flanders in September 1914. This left him with one leg shorter than the other.

Yet von Hünefeld, together with a German airline pilot and an Irish Army officer, became the first to make the North Atlantic crossing from east to west by aeroplane. Until then the only journey against the prevailing winds had been made by the British airship *R34* in July 1919.

Although none of the trans-Atlantic flights made during the 1920s was as celebrated as that of Charles Lindbergh, several other significant crossings were undertaken during the decade. They included that of von Hünefeld and his crew in the Junkers W 33 called *Bremen*, and also the first circumnavigation of the globe.

This had been accomplished in 1924 by a fleet of four specially built Douglas World Cruiser biplanes commissioned by the US Army Air Service that set off from Seattle to accomplish the feat. In the end only one managed to complete the journey, but it was still a great achievement even though at the time it attracted far less recognition than seemed due to its breathtaking scope.

The round the world flight obviously involved crossing the Atlantic, but such was the range of the ambitiously titled World Cruiser that the journey had to be made in a series of short hops. The aircraft was built for the sole purpose of attempting the first circumnavigation of the globe by air, but the itinerary had to be planned to avoid over-water journeys longer than 700 miles (1,120km).

The World Cruiser was a modified version of the DT-2 torpedo bomber that Douglas had built for the Navy. Its fuel system was completely redesigned to increase capacity from 115gal to 644gal (523lit to 2,928lit). Other changes included a different vertical tail with extra bracing struts beneath the tail and a smaller distance between the two open cockpits to improve communication between pilot and flight mechanic.

The 420hp Liberty V-12 liquid-cooled engine was retained, but radiator capacity was increased. The undercarriage was designed to be easily changed from wheels to floats depending on the need for operations from land or sea. The prototype, built in forty-five days at a cost of $23,721, was delivered to the Army for evaluation in November 1923. Soon after, Douglas received a contract for four production aircraft.

On 17 March 1924 four DWCs and their eight crew members left Santa Monica for Seattle, the official starting point. While there, the wheels were replaced by pontoon floats for the first stage of the journey over the Pacific, which began on 8 April when the aircraft, now renamed *Chicago*, *New Orleans*, *Boston* and *Seattle*, took off from Lake Washington. *Seattle*, however, was lost when it crashed into an Alaskan mountainside in thick fog and it was only after a ten-day trek through the frozen wilderness that the two-man crew reached Dutch Harbor. The other three World Cruisers continued. They were kept flying with the help of fifteen spare engines, fourteen extra sets of floats and duplicates of all airframe parts, stationed at various sites around the world.

By early August the remaining aircraft were now facing the Atlantic crossing, which was undertaken from the Orkney Islands via Iceland and Greenland to Nova Scotia. However, with nearly three-quarters of the flight completed, *Boston* made a forced landing in the Atlantic between the Shetland Islands and Iceland. The crew was rescued and two days later the prototype World Cruiser, now named *Boston II*, joined the group in Newfoundland to enable its original crew to complete the flight.

The fleet returned to Seattle on 28 September as landplanes. They had covered 27,553 miles (44,342km) in six months and six days in an actual flying time of 371 hours. They had touched down in twenty-eight countries and crossed the Atlantic and Pacific oceans. The flight was probably the greatest feat achieved by pioneer aviators up to that time. It certainly enabled the Douglas Aircraft Company to claim that its products had been 'First Around the World'.

Nevertheless, the perceived hazards of the trans-Atlantic crossing discouraged further attempts by pilots of heavier-than-air craft to attempt the journey, for a while at least. There was, indeed, no incentive, until Raymond Orteig offered his $25,000 prize for the first non-stop flight between New York and Paris.

Charles Lindbergh was not, however, the first to try and claim it. Several earlier attempts were made, notably by crews led by two of France's greatest First World War fighter aces, but in the end the Atlantic challenge defeated both of them.

Rene Fonck, who was credited with seventy-five wartime victories, received backing from Russian-born, American-domiciled aircraft designer Igor Sikorsky, who put up $100,000 and built an aircraft for the purpose. On 20 September 1926, the dapper Fonck and his crew of three attempted to take off from Paris, but crashed in the process. The big Sikorsky S-35 tri-motor, heavily laden with fuel, burst into flames and two crew members were killed. Fonck survived.

On 8 May 1927, Charles Nungesser (forty-three victories), took off from Paris with his navigator and wartime comrade Francois Coli to attempt the flight from east to west, something that until then had only been achieved by the British airship *R34*.

Nungesser and Coli's Levasseur PL-8 biplane, named *L'Oiseau Blanc,* disappeared en route and no trace of it or its crew was ever found. It was reported to have been sighted over Ireland, but the disappearance of Nungesser and Coli is considered to be one of the great mysteries in the history of aviation. It is thought that the aircraft was either lost over the Atlantic or it crashed in a remote area of Newfoundland or Maine.

The three other contenders for the Orteig Prize were all American: polar explorer Richard E. Byrd with Floyd Bennett and George Noville as crew, who had commissioned a tri-motor named *America* from Anthony Fokker; Clarence Chamberlin and Bert Acosta, who planned a flight aboard their Bellanca monoplane *Columbia*; and Stanton Wooster and Noel Davis in their Keystone Pathfinder named *American Legion*.

Byrd, however, crashed during a test flight that injured Bennett so seriously that he was unable to continue. His replacement was Bernt Balchen and, despite Lindbergh's success in claiming the $25,000 prize, he and Byrd took off for Paris on 29 June. But after a forty-hour flight they were unable to find Le Bourget and eventually landed at Ver-sur-Mer, Normandy, on 1 July.

Davis and Wooster both died when, on 26 April 1927, they crashed, also on a test flight. But Chamberlin actually managed to complete a pioneering flight of some significance. As part of their preparations, he and Acosta undertook a series of test flights in *Columbia*, increasing their aircraft's weight to test its capability. They also simulated the planned New York–Paris flight, setting an endurance record in the process, but arguments with the sponsor resulted in Acosta quitting to join Byrd. Chamberlin decided to make an attempt to claim a $15,000 prize put up by Brooklyn Chamber of Commerce for the first flight to Berlin.

Partnered by his sponsor, Charlie Levine, Chamberlin left New York on 4 June in *Columbia*. After a forty-two-hour flight they arrived over Germany, but were unable to find Berlin and, with fuel running low, were forced to land at Eisleben, 60 miles (96km) south-west of the capital. They finally reached Berlin on 7 June.

After the First World War, Ehrenfried Günther Freiherr von Hünefeld became a public relations man for the Norddeutscher Lloyd shipping line. Lindbergh's achievement had made him think that a successful east–west crossing could lead to regular two-way air travel linking Europe with North America.

The choice of equipment for his attempt to make the flight from east to west seemed obvious to the wealthy von Hünefeld. It would have to be a German machine and, given his friendship with aircraft manufacturer Hugo Junkers, he chose two of his new W 33 single engine, all-metal monoplanes. Von Hünefeld named them after Norddeutscher Lloyd's two flagships, the *Bremen* and the *Europa*. With additional fuel tanks and careful fuel management, he calculated that they should be able to fly thirty-eight hours non-stop.

Support for his venture came from Junkers himself and also from Herman Kohl, a First World War pilot and senior executive of Germany's airline, Deutsche Luft Hansa. As von Hünefeld was not a pilot, it was agreed that Kohl would actually fly the *Bremen*. He had joined the Army Air Service during the war and rose to command a bomber squadron. He also received Germany's highest decoration, the Pour le Merite, but a crash behind enemy lines resulted in his capture. He did, however, manage to escape and return to Germany. After the war he joined the national airline.

After a series of test flights in the *Bremen*, registered D-1167, during which they set new flight duration records, von Hünefeld and Köhl set off for Baldonnel, Ireland, from where they intended to start their trans-Atlantic flight. There they met James C. Fitzmaurice, the Irish Air Corps' airfield commandant, who agreed to accompany them on their flight.

James Michael Christopher Fitzmaurice was well-qualified to do so. Unlike his two companions, Fitzmaurice had some experience of flying over the Atlantic. The year before he had been part of a team that had attempted to cross the ocean, but the attempt was abandoned in the face of turbulent weather. After a flight of five hours thirty minutes the Fokker VIIa, called *Princess Xenia,* turned back when it was 300 miles (480km) from the Irish coast.

Irish-born Fitzmaurice had served with the British Army during the war and subsequently transferred to the RAF, although he had not completed his flying training before the Armistice. In May 1919 he was selected to undertake the first night mail flight, from Folkestone to Boulogne. After the creation of the Irish Free State he joined the new country's air service.

During their preparations for the flight, Kohl and Fitzmaurice calculated that it would take between thirty-three and thirty-six hours to reach New York. They expected to arrive in the early afternoon the day after a morning departure from Ireland. By mid-morning of the second day they expected to sight the coast of Newfoundland and then fly south along the US coast to New York.

The reality turned out to be somewhat different. The three men left Baldonnel in the *Bremen* at 05:38 hours GMT on 12 April. An hour and a half later they passed Slyne Head Lighthouse on the westernmost point of County Galway. Continuing west, they were soon out of sight of land. The weather was favourable, but when they passed the 30-degree line of longitude, they calculated that headwinds had reduced their speed to 90mph (144kph).

At 04:00 hours they climbed higher, hoping to improve their ground speed. As the sun set, they took their final drift sight reading. The three aviators hoped the winds would remain at the same strength overnight, although changes were inevitable as the hours ticked by. They expected to hold their course relative to the stars and based on their magnetic compass, but cloud was already obscuring the sunset and the *Bremen* was soon flying under a solid overcast.

After the sun went down, cloud obscured the stars, so the aircraft climbed to 6,000ft (1,830m) and held a solid course through the night based on the compass course alone. Before first light next morning, the overcast dissipated and the aviators had their first sight of the North Star. They realised that, instead of maintaining the correct course, they had been 40 degrees off it and angled northwards.

Köhl immediately turned south-west to make a course correction of nearly 90 degrees. As dawn broke, however, the aviators could see that the *Bremen* was now flying over land, but they could see nothing below them that corresponded to the features shown on their maps. All they knew was that they were somewhere over Canada.

Flying into wind meant that forward progress was slow. To avoid the worst of the wind, *Bremen* descended to lower altitude, but still no trace of human habitation could be seen. If their engine were to fail now the flyers might never be found, even if they survived the resulting crash-landing. With about two hours of fuel remaining the crew spotted a lighthouse on an island.

As they circled it the crew of the *Bremen* could see several four people and a pack of dogs. They decided to land and then discovered that they were at the Strait of Belle Isle, close to where the borders of three Canadian provinces, Quebec, Labrador and Newfoundland, meet. It was 18:10 hours GMT, about noon local time. They had flown non-stop for thirty-six hours thirty minutes.

When news of their achievement reached Europe and the United States, the three men were acclaimed as heroes. When they finally arrived in New

York they received a ticker tape welcome like that given to Lindbergh a year earlier. By a special act of the Congress the three aviators were awarded the Distinguished Flying Cross, which they received on 2 May from President Calvin Coolidge.

Six months later, the restless von Hünefeld hired a Swedish pilot to fly his second W 33, *Europa,* on an attempt to fly round the world. The aircraft reached Japan but during a stop caused by bad weather, von Hünefeld's health began to deteriorate and he abandoned the attempt and returned to Germany. Three and a half months later he succumbed to stomach cancer.

Hermann Köhl left Deutsche Luft Hansa and Berlin in 1935 to live on a farm in the country, but he died three years later. In contrast to his two companions on the *Bremen*, James Fitzmaurice lived a long and eventful life. He died in 1965 aged sixty-seven.

A year after Lindbergh's flight, Amy Guest, wife of prominent British politician Frederick Guest, expressed an interest in becoming the first woman to fly (or be flown) across the Atlantic. Guest was the daughter of American industrialist Henry Phipps and in 1905 she had married the man who would become chief whip in Lloyd George's wartime cabinet and secretary of state for air in the 1920s. She was well-known as a supporter of women's suffrage, a philanthropist and an aviation enthusiast, but decided against undertaking the flight herself. Instead she suggested that 'another girl with the right image' should be sought to take her place and offered to sponsor the project.

That girl turned out to be Amelia Mary Earhart. She had been born in July 1897 in Atchison, Kansas, to a lawyer of German descent and the daughter of a local bank president. Amelia was something of tomboy and grew up as an independent-minded feminist with an interest in mechanical things. She became hooked on flying after a chance visit to an airfield in California. 'As soon as I left the ground I knew I had to fly,' Earhart wrote later. She saved up for flying lessons and bought her first aircraft in January 1921. She used the bright yellow Kinner Airster to set a world altitude record for female pilots.

One afternoon in April 1928, Earhart took a telephone call in which she was asked: 'Would you like to fly the Atlantic?' The project co-ordinators, who included book publisher and publicist George P. Putnam – whom she later married – interviewed Earhart and asked her to accompany pilot Wilmer Stultz and co-pilot/mechanic Louis Gordon on the flight, nominally as a passenger, but with the added duty of keeping the flight log.

The team left Trepassey Harbor, Newfoundland, in a three-engine Fokker F.VIIb/3m on 17 June 1928 and landed twenty hours forty minutes later at Pwll near Burry Port in South Wales. A commemorative blue plaque marks the place where they landed. Since most of the flight was made on instruments and Earhart was not qualified for instrument flying, she did not take the controls. That hardly mattered, of course, because she had become the first woman to fly the Atlantic. But when interviewed after landing she was somewhat dismissive of her achievement. 'Stultz did all the flying – had to,' she said. 'I was just baggage, like a sack of potatoes.' She added: '... maybe someday I'll try it alone.'

The Junkers W 33

During the 1920s and '30s the Junkers company of Dessau, Germany, produced a series of advanced all-metal aerodynamically clean cantilever monoplanes, of which the W 33 was one.

Designed by Herman Pohlmann, it was developed from the four-seat F 13, which appeared in 1919. The W 33 featured an aluminium alloy duralumin structure covered with Junkers' characteristic corrugated dural skin with a large door on the port side giving access to the freight compartment. Power was provided by a 228kW Junkers L5 upright, in-line, water-cooled engine.

There was an enclosed two-seat cockpit and the undercarriage was fixed, although floats could be fitted to enable the aircraft to operate from water. The prototype, registered D-921, first flew as a seaplane, from Leopoldshafen on the River Elbe near Dessau in June 1926.

Production began in 1927 and continued until 1934, with 198 production machines built. Most were built at Dessau, although a few were assembled at Junker's Swedish subsidiary AB Flygindustri near Malmo and in the USSR. The Swedish plant had been set up in the early 1920s to avoid post-war restrictions on aircraft construction in Germany, while the Russian works at Fili near Moscow was primarily used for fighters for the Red Army. There were more than thirty W 33 variants.

The type was used for several significant flights in addition to that of the *Bremen*. That aircraft is now on display at Bremen Airport, Germany. Another W 33 set world records for both endurance and distance on a single flight from Dessau. Johann Rustics and Cornelius Edzard covered 2,896 miles (4,661km) in fifty-two hours twenty-two minutes between 3 and 5 August 1927. A little earlier, Fritz Loose and W.N. Schnabele had set another record for duration and distance, but this time carrying a 500kg load. They stayed aloft for twenty-two hours eleven minutes, during which time they travelled 1,701 miles (2,736km). A W 33 with floats set similar records for seaplanes. On 26 May 1929, a much-modified W 33, also referred to as a W 34, established a world altitude record of 41,800ft (12,740m) piloted by Willy Neuenhofen.

The first Swedish-assembled W33 was delivered to Mitsubishi in Japan and in 1932 was used for an attempt to fly across the Pacific Ocean to the US. But the attempt failed, the aircraft disappeared and, despite a search that lasted six months, no trace of it was ever found.

Junkers W 33 Specification

Accommodation	2–3
Length	10.50m (34ft 5.5in)
Wingspan	17.75m (58ft 2.75in)
Height	3.53m (11ft 7in)
Max take-off weight	2,500kg (5,511lb)
Power plant	Junkers L 5 in-line engine developing 310hp (228kW)
Maximum speed	120mph (180kph)
Range	620 miles (1,000km)
Service ceiling	14,100ft (4,300m)

Chapter Nine

Have you Flown Far?

Dieudonne Costes cut a dashing figure posing in his flying gear beside the bright red biplane emblazoned with the names of the faraway places he had visited and decorated with the stork insignia of one of France's most famous fighter squadrons.

The man whose aviation achievements would continue to be celebrated during the late 1920s and early '30s, had indeed served as a fighter pilot during the First World War, being credited with six confirmed victories.

It was two years after the *Bremen*'s flight that Costes (some accounts omit the final 's') and his navigator, Maurice Bellonte, succeeded in reaching New York from Paris. According to the *New York Times*, the 'dramatic reciprocal gesture' to Lindbergh's flight made instant heroes of the pair both in France and the USA. The paper observed that their flight also signified that 'not only France and America but other countries, too, look forward to the day when regular air travel across the ocean will be commonplace'.

Costes and Bellonte met after the First World War when they were working for the French airline Air Union. They often flew on the Paris–London route together. Costes had completed a number of long-distance flights before he and Bellonte decided to attempt the New York trip. Between October 1927 and April 1928 he and Joseph Le Brix flew 35,652 miles (57,410km) around the world in a Breguet 19GR called *Nungessor-Coli*.

It was during this trip, on 14–15 October, that they made the first non-stop aerial crossing of the South Atlantic Ocean, flying between Senegal and Brazil. They subsequently flew through every country in South America. On 15–17 December 1928 Costes set a world closed circuit distance record of 8,029km (4,986 miles).

Like the other successful trans-Atlantic aviators, Costes and Bellonte prepared for their flight with great thoroughness. In fact, their preparations took several years as they tested equipment and established several records before committing themselves to the trans-Atlantic flight. In 1929 the pair established a long-distance record by flying the 4,910 miles from Paris to the

Manchurian city of Quqihar non-stop in their Breguet 19 *Point d'Interrogation*. In November they made the first flight from Paris to Hanoi.

By early 1930, Costes and Bellonte had postponed their Paris–New York flight several times because of bad weather. On one occasion they actually took off but were forced to turn back and it was not until 1 September that they decided conditions were right. Lindbergh had flown alone aided by just a few rudimentary instruments and without radio, but Costes and Bellonte used more than thirty navigational instruments, some of which were duplicated for safety's sake. They also had one for maintaining a true course in fog as well as a radio.

After taking off from Paris in foggy conditions, the aviators had to dodge several storms, but otherwise enjoyed relatively favourable flying conditions for more than thirty-seven hours before they landed safely at Curtiss Field, Valley Stream, Long Island. They were subsequently treated to a ticker tape parade through Manhattan and were entertained by President Hoover at the White House. They also completed a twenty-five-day goodwill flying tour of the USA.

Both aviators went on to have long and eventful lives, Costes serving as a flying instructor in the Second World War, while Bellonte was in the Resistance. Costes died in 1973 aged eighty, while his colleague, who had become Inspector General for Flight Safety with the French government, died in 1994 aged eighty-seven.

Two years after Costes and Bellonte's flight, Jim Mollison made the first east–west solo crossing of the Atlantic. Mollison, who had earlier been the RAF's youngest officer and also the most youthful instructor at the Central Flying School, was already well-known for his long-distance flights. In August 1932 he flew from Portmarnock Strand near Dublin to Pensfold, New Brunswick, in a single-engine de Havilland Puss Moth registered G-ABXY and named *The Heart's Content*.

The following February, Mollison used the same aircraft to make the first east–west solo crossing of the South Atlantic, from Lympne in Kent to Natal, Brazil. Soon after his first trans-Atlantic flight Mollison met and married Amy Johnson, who, a year earlier had become the first woman to fly solo from England to Australia. The two made several long-distance flights together but divorced in 1938.

Despite the achievements of Mollison and others, the airship was still seen as the vehicle for long-distance aviation, especially as in 1929 the *Graf Zeppelin*, commanded by Hugo Eckener, had completed a circumnavigation of the globe in twenty-one days.

One man who had a burning desire to beat this was Oklahoma-domiciled Texan Wiley Post. After a somewhat chequered career, during which he had lost an eye in an accident, Post had become the personal pilot of Oklahoma oil magnate F.C. Hall. In 1930 Hall bought a high-wing, single-engine Lockheed Vega, which he named *Winnie Mae* after his daughter. Post achieved national prominence when he used it to win the National Air Race Derby from Los Angeles to Chicago.

In the autumn of 1930, Post would write later, cross-country flying had become monotonous and the 'steady grind' of piloting Hall from one bad landing field surrounded by oil derricks to another was losing its appeal. And besides, the effects of the Great Depression were beginning to encroach into the oil business and delay Hall's expansion plans.

Post and his employer had an understanding that he would be free to make special flights 'in the interests of aviation or for my own amusement'. Despite the impetus of Lindbergh's flight, the public, Post felt, still needed more definite proof of the aeroplane's reliability under all conditions. By the early 1930s many of the great aviation journeys had been completed so what better than a flight around the world to demonstrate the supremacy of the aeroplane over the airship?

Hall allowed Post the use of the *Winnie Mae* and offered financial backing to add to the few thousand dollars his pilot had saved. Post also considered he had the right experience to undertake the flight and knew he could count on help from Australian-born former merchant seaman Harold Gatty, who was running a navigation school in California. Gatty's input had helped Post win the Los Angeles–Chicago race. 'His charts,' Post would write, 'were so accurate that even after my compass broke when I was half-way through the long run, I had held the 'plane true and beat my nearest rival by more than 30 min in a 10 hours flight.'

Gatty agreed to accompany Post and on 31 January 1931 *Winnie Mae* headed for California and the Lockheed and Pratt and Whitney factories to be prepared for the round-the-world flight. Post calculated that he would need at least 500gal (2,270lit) of fuel for the longest hop of 2,550 miles (4,080km). An extra fuel tank was installed in the fuselage behind the pilot. Gatty's station was aft of the tank with all his instruments, radio and folding table. There was a hatch in the cabin roof through which he could take sightings of the sun and stars while protected by a folding windscreen. Pilot and navigator would communicate via a speaking tube.

The month-long wait for favourable weather at Roosevelt Field, New York, came as an anti-climax but finally, on 23 June, the *Winnie Mae* splashed through the rain to begin the circumnavigation. It was 04:55 local time; 1,153 miles (1,845km) and six hours forty-seven minutes later the blunt-nosed Vega was landing at Harbour Grace, Newfoundland.

Post and Gatty were off again after lunch, but this time the Atlantic lay ahead. By the halfway point they ran into fog so thick that, as Gatty wrote later, 'I now know what it means to be in solitary confinement.' Post tried to fly around the fog, but it was so extensive he had to give up and return to the original course. After a brief respite they found themselves boring through cloud again. By this time Gatty calculated that they should be over land and Post put the aircraft into a shallow dive. They emerged with 1,500ft (450m) showing on the altimeter. 'Water!' Gatty shouted. 'Water, hell,' Post replied, 'land!'

Just twenty minutes later they sighted an airfield. Post would later recall the thrill that the subsequent touch-down gave him, even though he and Gatty were not entirely sure where they were. They soon learned that they had landed at RAF Sealand near Chester. But there would be another slight delay. The officers, Post wrote later, 'wouldn't hear of our leaving without lunch. We had some good old English roast beef like that advertised by New York chop houses.' He added: 'The fellows from the RAF were so nice that we hated to leave, but naturally we were anxious to get on our way as soon as possible.'

The *Winnie Mae* needed 500ft (150m) of Sealand's runway to get airborne and was soon crossing the coast at Lowestoft en route for the next landing at Berlin 700 miles (1,120km) away. From there the *Winnie Mae* flew the 1,000 miles (1,600km) to Moscow and on the fourth day made the 2,000-mile flight to Novosibirsk in Siberia. After an eight-hour stop-over Post and Gatty were off again, to Irkutsk, arriving there on the fifth day. Over the next two days *Winnie Mae* continued to work her way around Siberia before attempting what was considered the most hazardous part of the whole trip, the 2,500-mile (4,000km) flight across Kamchatka and the Bering Strait to Solomon, Alaska.

They arrived there on the eighth day and then departed for Fairbanks 500 miles away. By 1 July the Vega had arrived at Edmonton for the last 2,000 miles of the round trip back to New York via Cleveland. Post and Gatty had covered 15,474 miles (24,903km) in the record time of eight days, fifteen hours and fifty-one minutes.

Everywhere they went the reception they received rivalled that given to Charles Lindbergh. They had lunch at the White House on 6 July, rode through New York City in a ticker tape parade the next day and were guests of honour at a banquet given by the Aeronautical Chamber of Commerce of America. After the flight, Post acquired the *Winnie Mae* from F.C. Hall. 'I had satisfied my life's ambition,' he said. But two years later he decided to try and beat his time and, moreover, to do it flying solo.

This time he flew direct from New York to Berlin, covering the 3,600 miles (5,760km) in twenty-three hours forty-six minutes. He left New York on 16 July 1933 to complete what would be the first non-stop flight to the German capital. From there he continued via Konigsberg, Moscow, Novosibirsk,

Irkutsk, Rukhlova, Khabarovsk, Flat (Alaska), Fairbanks, Edmonton and back to New York. The flight took him seven days, eighteen hours and fifty minutes.

The same month a group of twenty-four Italian flying boats reached Chicago, having flown in formation from Orbello, near Rome. They were led by General Italo Balbo, a crony of Italian Fascist leader Benito Mussolini. Edwin Shepherd, writing in 1939, said that the 'flamboyance' of the flight, designed to demonstrate the prowess of Italy and its reorganised air force, 'could not overshadow the excellent work done in arranging it and in bringing a big formation safely over thousands of miles of the earth's surface'.

Savoia-Marchetti S.55s had already flown the South Atlantic and now the intention was to fly a formation of twenty-five of the twin-hulled flying boats to Chicago, where the World's Fair was being held. Compared with the standard S.55s then in service, Balbo's aircraft were fitted with Isotta Fraschini engines offering more than 50 per cent more power than the original Fiats. The power plants were mounted back-to-back on pylons over the wing behind the cockpit. The cockpit represented the thickest part of the wing leading edge. The rest of the crew, passengers and cargo travelled in the two hulls mounted under the wings and linked to the tail by struts and wires.

For the Chicago flight the S.55s carried a crew of four and included a wireless operator and flight engineer. They took off on 1 July 1933 but one was lost, together with a crew member, on the first leg to Amsterdam, when it hit the edge of a dyke on landing. The following day the remaining twenty-four aircraft flew to Londonderry, Ireland, and three days later the weather was considered favourable for the next leg, to Reykjavik. The aircraft encountered storms and heavy rain, but reached their destination safely.

There was a further delay caused by the weather but on the 11th the aircraft, now heavily loaded with fuel, attempted to take off. All but Balbo's managed it. Another attempt, the following day was successful and the formation set out to cover the 1,520 miles (2,430km) to Cartwright, Labrador. The aircraft became separated in the fog and it was twelve hours before all were safely down at their destination. From there they flew to Montreal via Shediac, New Brunswick, and then on to Chicago.

Although the weather had made formation-keeping difficult, the twenty-four S.55s made a spectacular arrival over Chicago. Over the next four days Balbo and his crews were feted at dozens of ceremonies. The general, however, was anxious to be back home by the end of the month and the formation left for New York on the 19th. There were more festivities and Balbo and some of his crews visited Washington. By the 29th they were back in Newfoundland and ready for the crossing to Ireland, but bad weather indicated that it might be better to go via the Azores.

Just before dawn on 8 August they took off and twelve hours later had covered the 1,200 miles (1,920km) to Horta and Ponta Delagada, where the aircraft were moored for the night. The next day one was damaged while attempting to take off, blocking the way for three others. Balbo decided to push on without them and the twenty survivors reached Rome on the 12th. Edwin Shepherd observed that the trip had broken no records but that 'its merit lay in the scale of which it had been done'. One legacy, however, was that a large formation of aircraft became known as 'a Balbo'.

By that time Amelia Earhart had already redeemed the pledge she made in 1928 to become the first woman to fly the Atlantic solo. The Irish farm worker who saw the little bright red high-wing aircraft buzz out of the western sky and come safely to rest in his field certainly did not know what to make of the slim figure with grease-stained face who emerged from it. 'I couldn't tell if it was a man or woman,' Dan McCallon would later tell reporters. He asked the traveller: 'Have you flown far?' The reply was brief but enough to put McCallon right: 'From America.'

He recalled: 'She answered calm, like. I was stunned and didn't know what to say.' But Earhart was already a heroine in her own country. Now the woman they called 'Lady Lindy' had truly emulated Charles Lindbergh to become the first woman to make the crossing solo.

Her intention when she left Harbour Grace, Newfoundland, on 20 May 1932 had been, like Lindbergh, to fly to Paris. But, carrying little more than a toothbrush, some hot soup, three cans of tomato juice and a copy of that day's *Telegraph-Journal* newspaper, she had been forced to land at Culmore near Londonderry, Northern Ireland. Earhart's journey had been cut short by a string of troubles: a broken altimeter, cracked engine exhaust manifold, a fuel leak and iced-up wings.

Fighting fatigue, Earhart had realised that Paris was out of the question. In any case, turning back would mean a dangerous night landing on an unlit airfield. As she drily noted later: 'One crash is as good as another, so I kept right on going.'

And she did. The 2,026-mile (3,240km) flight, which took just four minutes under fifteen hours to complete, wrote Earhart's name in the history books. There were, of course, no reporters to witness this triumph, but in Londonderry she was greeted by the city's mayor. 'I have,' Earhart declared, 'realised my ambition.' New York gave her the inevitable ticker tape welcome and she was feted worldwide.

Earhart would complete several more spectacular long-distance flights, but her real ambition was to fly around the world. It was while attempting that feat in July 1937 in a twin-engine Lockheed 10-E Electra that Earhart and her navigator, Fred Noonan, disappeared to create one of the greatest mysteries of the twentieth century.

Ironically, their disappearance was to lead indirectly to one of the most bizarre flights in the history of aviation. It might have sounded like a music hall joke yet there really was an Irishman who had planned to fly to California, but ended up in Ireland instead.

He was apparently intending to fly from New York to Long Beach, but misread his compass. Actually, though, it was probably Douglas Corrigan who had the last laugh. It was he who put about the story that he'd mistakenly flown a reciprocal bearing at a time when the authorities were preventing him from attempting the trans-Atlantic crossing in his ramshackle single-engine aircraft.

Whatever the truth, the man who had briefly been involved with Ryan and Charles Lindbergh at San Diego would be known for the rest of his life as 'Wrong Way' Corrigan. And it would be quite a long life, too. He died in 1995 aged eighty-eight having shrewdly capitalised on his fame with product endorsements – including a watch that ran backwards – an autobiography and a Hollywood film in which he starred as himself. He returned to the USA to a ticker tape welcome, a meeting with the president and a triumphal nationwide tour.

Of Irish descent, Corrigan was originally called Clyde after his father, but changed his name when his father abandoned his mother and three small children. He had learned to fly in 1926, paying for lessons by going without lunch. Over the next two decades he would have a variety of aviation jobs, such as mechanic, assembly worker and barnstorming pilot. In 1927 he was employed by Ryan at San Diego and helped assemble the *Spirit of St Louis*.

He left Ryan briefly, but by 1935 he was back in what he now considered a dead-end job. What Corrigan really wanted to do was refurbish his four-year-old Curtiss Robin cabin monoplane and install a more powerful engine and additional fuel tanks. His objective was to 'try to fly from Newfoundland to Ireland'.

Corrigan bought a used Wright J6-5 five-cylinder air-cooled radial engine and had it shipped from New York to California. He also increased fuel capacity to 225gal (1,023lit). But when he sought official permission to make to the flight his application was rejected on the grounds that the Robin was unsuitable for a non-stop trans-Atlantic trip. It was, though, certified to a lower standard for cross-country journeys.

Over the next two years Corrigan made more modifications. He also made repeated filings for full certification, but none succeeded. Indeed, by 1937, the authorities refused to renew the Robin's airworthiness licence despite extensive modifications in the face of increasing regulation that had taken Corrigan's total outlay on the aircraft to more than $900. The authorities considered it unstable and unsafe. Corrigan believed the decision was at least partly due to increased official caution following the disappearance of Earhart and Noonan.

In his autobiography, *That's My Story*, Corrigan expressed his exasperation with this official resistance. Indeed, he is widely thought to have decided to make the crossing without permission, although he never admitted it. Corrigan gained an experimental licence and obtained permission for a transcontinental flight to New York with conditional consent for a return trip to Long Beach. He arrived at Floyd Bennett Field, New York, unannounced and virtually unnoticed due to the impending departure on a round-the-world flight of Howard Hughes.

After a twenty-seven-hour trans-continental flight Corrigan arrived on the east coast with a fuel leak but decided that repairing it would take too long. His logged flight plan had him returning to California on 17 July. Corrigan asked the manager of Floyd Bennett Field, Kenneth P. Behr, which runway he should use. Behr told him to use any one as long as he did not take off to the west, in the direction of the administration building where Behr had his office.

According to Corrigan's autobiography, Behr told him: 'I won't say good-bye, I'll say *bon Voyage*,' perhaps guessing at Corrigan's plan to fly the Atlantic. He would, however, later swear that he no idea what Corrigan was really up to as, at 05:15 hours on 18 July 1938, the Robin took off, its tanks brimming with 320gal (1,200lit) of fuel and 16gal (61lit) of oil. Once airborne, Corrigan headed east and disappeared into cloud.

He later claimed that one of his two compasses was not functioning and that he had set his westerly course on another, mounted on the cockpit floor. 'I turned the 'plane until the parallel lines matched and flew on over the fog,' he would write. He also maintained that for the first two hours he could see little through the fog until he caught sight of a city which he took to be Baltimore. 'I have since found out it was Boston, of course,' he said afterwards.

Eight hours from New York the clouds above and below became solid. Two hours later Corrigan became aware that his feet were cold. Looking down he saw why: the fuel leak had become worse and petrol was sloshing around the cockpit floor. The leak increased during the night so that the beam of his torch showed it was an inch (2.54cm) deep. To avoid fuel leaking out and igniting on the hot exhaust pipe, Corrigan punched holes in the opposite sides of the floor with a screwdriver.

That, however, meant that his plan to run the engine slowly to conserve fuel would have to be abandoned: the slower he flew the more time the tank would have to drain. He therefore increased engine speed from his planned 1,600rpm to 1,900rpm.

He continued to fly on instruments until 'I came down out of the clouds and saw nothing but water underneath.' He wrote later: 'This was strange as

I had only been flying 26 hours and shouldn't have come to the Pacific Ocean yet.' It was then, he claimed, that he discovered his mistake. 'I looked down at the compass and now that there was more light, I noticed I had been following the wrong end of the magnetic needle on the whole flight. As the opposite of west is east I realised that I was over the Atlantic Ocean somewhere. But where?'

It was at this point that Corrigan realised he was hungry and decided to dig out his provisions for the journey: two boxes of fig bars and two chocolate bars. Just then he sighted green hills below him and forty-five minutes later reached another coast, which convinced him he was over Ireland. After a further thirty-five minutes he sighted a large town, which he realised was Dublin. He landed at Baldonnel aerodrome twenty-eight hours thirteen minutes after leaving New York.

The first man he encountered on the ground was an Irish army officer. 'My name's Corrigan,' he said. 'I left New York yesterday morning headed for California, but I got mixed up in the clouds and must have flown the wrong way.'

Later, he was questioned by Mr Cudahy, the US diplomatic representative, who told him: 'You seem to have arrived unexpectedly. What caused the trouble?' Corrigan replied that it was hazy when he left New York. 'Well,' Cudahy responded, 'your story seems a little hazy too – now come on and tell me the real story.'

Corrigan insisted his account was the real story, to which Cudahy responded by asking if he was sticking to it. 'That's my story,' Corrigan insisted, 'but I sure am ashamed of that navigation.'

Whatever the truth of his story, Corrigan had demonstrated that the pioneering phase of trans-Atlantic flight was nearing its end. Technological advances had made aircraft, meteorology, and wireless communications more sophisticated and reliable than in the past. As long-distance flight became increasingly safe and commonplace, government and industry were expressing interest in establishing regular passenger and mail services.

The Wooden Bullets

The high-wing single-engine Vega, which flew for the first time in 1927, was the first product of the recently created Lockheed Aircraft Corporation. It also introduced what was to become the company's trademark practice of giving its products astral names.

Designers John Northrop and Gerard Vultee, both of whom would later form their own companies, had intended to create a four-seat aircraft that was not only rugged but also fast. This was to be achieved with a wooden monocoque fuselage, plywood-covered cantilever wings and a powerful engine.

Construction involved a method that had been patented by Northrop and Allan Lockheed. This involved building up the fuselage in two halves with laminated sheets of spruce glued together in a concrete mould. A rubber bladder was lowered into the mould and inflated with air to compress the wooden construction into shape against the inside of the mould. The two halves were then nailed and glued over a separately constructed framework.

The single-spar cantilever wing had to be mounted on top of the fuselage. Only the engine and landing gear remained essentially unstreamlined, although production aircraft featured teardrop-shaped 'spats' over the wheels.

The prototype (2788) made its first flight on 4 July 1927 from a field at Inglewood, now Los Angeles International Airport. The aircraft had already been sold, at a loss, to San Francisco newspaper owner George Hearst Jr, who entered it in the California–Hawaii air race. Equipped with additional fuel tankage, rapid-inflating flotation bags and an undercarriage that could be jettisoned in the event of a ditching, the aircraft was christened *Golden Eagle* and re-registered NX913.

Pilot Jack Frost and navigator Gordon Scott established a number of point-to-point records while preparing for the race and showed the aircraft capable of reaching 135mph (217kph). In doing so they established themselves as firm favourites. They took off from Oakland on 16 August, but were never to be seen again. Several other competitors were lost in the event.

Yet orders for the Vega rolled in, resulting in the construction of another 128 examples with numerous variants and one-off specials. The first production version (twenty-eight built) was the Vega 1, which could carry four passengers. The Whirlwind engine was uncovered, but later variants featured NACA cowlings that completely enclosed the Pratt and Whitney Wasp engine.

The 'Wooden Bullet' could fly at up to 185mph (298kph) and this, combined with predictable handling and durability, enabled it to stand out among its late 1920s contemporaries.

Lockheed Vega 5 Specification

Accommodation	pilot and six passengers
Length	28ft 6in (8.7m)
Wingspan	41ft (12.5m)
Height	6ft 5in (2.0m)
Loaded weight	4,500lb (2,041kg)
Power plant	Pratt and Whitney R1340C Wasp nine-cylinder air-cooled radial developing 450hp (335.6kW)
Cruising speed	165mph (265kph)
Range	725 miles (1,165km)
Service ceiling	25,000ft (9,570m)

Chapter Ten

The Silver Fish

Imagine a flying luxury hotel able to cross the Atlantic in less than half the time taken by the fastest ocean liner and you will have some idea of what it was like to travel in the last and greatest of the German airships.

During the mid-1930s these craft offered a unique form of intercontinental travel that seems far removed from anything available to twenty-first century flyers. Every departure was an event.

A brass band played as passengers presented their passports for inspection. Before boarding they were asked to surrender cigarette lighters, matches and even camera flashbulbs, anything that might risk igniting the highly flammable hydrogen gas that provided the lift.

At the appropriate time Captain Max Pruss ordered '*Schiff Hoch*!' The ground crew let go of the mooring ropes. Water ballast was released and the ship rose majestically into the air. The big diesels were then started to drive the ship towards its destination far across the ocean.

There was more ceremony to come at meal times. The silk walls of the elegant and spacious dining room were tastefully decorated in a contemporary style. The tables were covered in crisp white linen cloths with vases of flowers. The food was prepared by traditionally attired top-quality chefs and served on monogramed china by white-coated stewards. There was a choice of French or German wines.

In between meals, passengers could relax in the lounge or gaze down from the promenade deck as the world passed by hundreds of feet below at 85mph (136kph). Passengers could tour the craft led by the ship's doctor while their children were looked after by stewardess Emmi Imhof.

And, despite the 7 million cu ft of flammable gas within the gigantic envelope, there was a smoking room. It was sealed by an airlock and a special electric lighter was provided for smokers to use. In the evenings there were recitals on the special lightweight grand piano – not carried on the last voyage – after which passengers retired to bed in their individual cabins. They could even leave their shoes in the corridor to be cleaned.

There was no doubt about it: this was the way to travel. It was certainly the most comfortable and the swiftest way of crossing the Atlantic between Europe and North and South America, but at a price. A one-way trip from Germany to New Jersey cost as much as a family car. Small wonder that the *Hindenburg*'s passenger lists were studded with the names of celebrities as well as the era's wealthiest and most powerful figures.

Inevitably, history has viewed the German airships through the prism of the *Hindenburg*'s last flaming moments, to say nothing of their unsavoury association with the Nazis, who exploited their propaganda value. But nothing can change the fact that for a few brief years they offered a unique link between the Old World and the New.

The *Graf Zeppelin* was the first aircraft to provide regular intercontinental commercial passenger services when it began operations to Latin America in 1932. Between then and May 1937, when the service was abruptly withdrawn the day after the *Hindenburg* disaster, *Graf Zeppelin* flew the South Atlantic 136 times.

The man who could be called the father of the commercial airship did not live to see the success of the craft named after him, nor witness the tragedy of the ship that up to then represented the apotheosis of light-than-air technology.

Ferdinand Adolf August Heinrich Graf von Zeppelin was born in 1838 and died in 1917. He attended the military academy at Ludwigsburg, near Stuttgart, and at the age of twenty became an officer in the army of Württemberg. He gained his first flying experience as an observer during the American Civil War. Carrying a pass signed by President Abraham Lincoln enabled him to travel with the northern armies, but it was in St Paul, Minnesota, far from the war zone, that Zeppelin caught sight of a balloon. It would alter the whole course of his life and many others.

After making his first balloon flight, Zeppelin became convinced of aviation's great military potential and he conceived the idea of a rigid-framed steerable airship. It would have a metal framework comprising rings and longitudinal girders with individual cells containing hydrogen. In 1887 Zeppelin formally proposed the use of airships for military purposes, but it was not until after his early retirement from the Army in 1890 that he was able to devote himself to lighter-than-air flight. Within ten years he would build his first airship, *Luftschiff Zeppelin 1* or *LZ-1*.

By August 1914 and the outbreak of the First World War the German armed forces had several Zeppelin airships, each capable of travelling at about 85mph and carrying up to 2 tons of bombs. The military deadlock on the Western Front prompted the German High Command to use these craft to attack British towns and cities. Nine people were killed and buildings damaged in the first raid, on Great Yarmouth and King's Lynn, in January 1915. The impact on a population brought face to face with the realisation that the battlefield had suddenly moved to their own backyards was profound. There were further raids on coastal towns during 1915 before London became the target in 1916. The stealthy attackers arrived without warning so that, in the absence of purpose-built shelters, people had to hide in cellars or under tables.

There seemed to be no defence against these silent killers. Morale dropped as people feared further raids and even a German invasion. But in May 1916 British fighter aircraft were armed with guns firing incendiary bullets, which ignited the hydrogen gas and turned the attackers into flaming death traps. The menace had been overcome but more than 500 people had died in fifty-two attacks. To the British the name Zeppelin was now synonymous with death and destruction.

Although pleased by the military contribution his airships had made, Zeppelin had little personal involvement with them during the war. He had turned his attention to other engineering projects, including heavier-than-air craft and giant bombers. He died at the age of seventy-eight, but not before his creations had also found a more peaceful use. The world's first passenger airline, DELAG (*Deutsche Luftchiffarts-Aktiengesellschaft* or German Airship Transportation Corporation Ltd), was established on 16 November 1909 as an offshoot of the original Zeppelin company.

Between 1910 and the outbreak of the First World War, DELAG's Zeppelins carried more than 34,000 passengers on more than 1,500 flights without a single injury. Although most of the passengers had been members of German royalty, military officers, aristocrats, government officials and business leaders who received free flights to publicise the Zeppelin industry, DELAG also carried 10,197 paying passengers.

Of the company's first half dozen craft, three were lost in crashes and the remainder impressed into military service. After the war a new ship, *LZ-120*, was built for fast air transportation between Friedrichshafen and Berlin. Construction was completed within six months and the ship, named *Bodensee*, made its first flight in August 1919.

Its teardrop shape was derived from wind tunnel testing. It differed substantially from the thin, pencil-like outline of most previous craft and, as a result, offered less drag and greater aerodynamic lift. With its four 245hp Maybach MB IVa engines it could reach a speed of 82mph (131kph). Bodensee represented a big leap forward in design and became the basic model on which *LZ-127 Graf Zeppelin* and *LZ-129 Hindenburg* were based.

With washrooms and a small kitchen provided for the preparation of light meals, *Bodensee* could carry up to twenty-six passengers in comfort as well as speed, and in 1919 began scheduled services between Berlin and southern

Germany, taking between four and nine hours, compared with eighteen to twenty-four by rail. In the three months following the airship's launch, *LZ-120* made 103 flights and carried almost 2,500 passengers, 11,000lb (4,900kg) of mail and 6,600lb (3,000kg) of cargo.

DELAG acquired a second ship from the Zeppelin company in 1920. *LZ-121 Nordstern* was intended to operate international passenger services between Friedrichshafen, Berlin, and Stockholm. However, it had not entered service when DELAG was forced to cease operations under the terms of the Treaty of Versailles. The two airships were transferred to the Allies as war reparations, *Bodensee* to Italy, and *Nordstern* to France.

After the war the Allies placed harsh restrictions on all German aviation activities, particularly those involving lighter-than-air craft. There were limits on the size of airships that could be built. The Zeppelin company was effectively put out of business, making the construction of a new intercontinental airship highly unlikely. Or it would have been but for the efforts of Hugo Eckener. He was an early associate of Count Zeppelin, earning fame as an effective airship commander and tireless advocate of light-than-air passenger craft, deploying a combination of technical skill, business acumen and the ability to inspire enthusiasm.

Eckener was born in 1868 and later gained a doctorate psychology. But it was while he was working as a journalist for the *Frankfurter Zeitung* that he was assigned to cover the second flight of *LZ-1* in October 1900. From then on, he became a devoted disciple of the count, initially working for his company as a public relations man.

Before long, however, Eckener had become deeply involved in the technical and operational aspects of airships and by 1911 was given his first command. His first flight, in charge of *LZ-8 Deutschland II*, ended somewhat ignominiously. In May 1911 the craft was being moved from its hangar when a gust of wind tore the ship from its ground crew and smashed it against the hangar. There were no injuries, but the passengers and crew had to be rescued by a long fire ladder.

By this time Eckener had established a reputation as a strict, almost severe, commander who had little patience for incompetence or lack of effort. He was a formidable, imposing personality who stood on ceremony and expected to be addressed formally and never by his Christian name. On the other hand, he was also known for his fairness to subordinates and his calmness during moments of crisis and tension. As a commander he was respected for his ability to remain on the bridge for days at a time when conditions warranted it.

Although he remained a civilian during the war, Eckener was deeply involved in Germany's deployment of airships as weapons. As the Navy's director of airship training, he was responsible for the instruction of more than 1,000 airship crewmen. But it was after the war that Eckener made what was probably his greatest contribution to the Zeppelin project.

The reparations Germany was required to make included the construction of an airship for the Americans. Eckener convinced the Allied authorities to allow the Zeppelin company to build a new ship, *LZ-126*, which would be delivered to the US Navy as the USS *Los Angeles*.

This kept the Zeppelin company alive, maintaining not only its plant and equipment, but also its highly skilled workforce. The construction and operation of LZ-*126* also gave Eckener and his colleagues the knowledge and experience they would use in building *Graf Zeppelin* and *Hindenburg*.

Under Eckener's command, *LZ-126* left Friedrichshafen on 12 October 1924 and arrived at the US Navy base at Lakehurst, New Jersey, at 09:56 hours on the 15th, having covered just over 5,000 miles (8,000km). It was a triumph for Eckener and the Zeppelin company. For the first time, mail and freight from Berlin had reached Manhattan in less than five days. Far from being treated as recent enemies, Eckener and his crew were welcomed as heroes. After a ticker tape parade along New York's Broadway they were welcomed at the White House by President Calvin Coolidge.

As the British had opposed the construction of a new Zeppelin, it was agreed that the ship would be for civil rather than military use. As a result, it was given a level of accommodation suitable for a long-distance commercial airliner with a large passenger cabin featuring sleeping compartments and a first-class galley for the preparation of hot meals. The Navy operated *Los Angeles* primarily as a training ship, although it did make several notable long-distance flights.

Encouraged by this success, Eckener decided the time was right for a new Zeppelin. Of the total cost of 7 million marks, 2.5 million was raised by a lecture tour by Eckener and his senior colleagues. But an additional million marks from the government combined with finance drawn from other operations enabled the Zeppelin company to begin construction. Eckener was very impressed by the completed craft when he first saw it. He thought it looked like 'a fabulous silver fish, floating quietly in the ocean of air and captivating the eye just like a fantastic, exotic fish seen in an aquarium'.

LZ-127, later named *Graf Zeppelin*, was to become the most successful craft of its type, completing more than a million miles on 590 flights and carrying more than 34,000 passengers without a single injury. It completed many significant and pioneering flights under Eckener's command, including trans-Atlantic crossings, a round-the-world flight and an exploration of the North Pole. In addition to demonstrating the viability of airship travel, these feats won Eckener international fame. *Time* magazine featured him on its front cover in 1929.

The new ship was named *Graf Zeppelin* by Zeppelin's daughter in July 1928 and made a three-hour maiden first flight in September commanded by Eckener. A series of tests followed, including a thirty-four-and-a-half-hour endurance flight, during which the new airship was demonstrated to the residents of thirteen German cities including Eckener's home town of Flensburg.

After domestic shake-down flights the craft made its first long distance journey, to the USA, later in 1928. Later flights included a round-the-world tour in August 1929, a pan-American flight in 1930, a polar expedition in 1931, two round trips to the Middle East and a variety of other flights around Europe.

Less than a month after its first outing, the *Graf Zeppelin* made the first commercial passenger flight across the Atlantic. On board were forty crew, commanded by Eckener, and twenty passengers, who included a US Navy officer and a reporter from Randolph Hearst's group of newspapers, Lady Grace Drummond-Hay. The widow of a British diplomat, Drummond-Hay brought a touch of glamour to the event as one of half a dozen reporters covering the flight. But as the only woman on board she inevitably became part of the story, receiving a great deal of attention from the world's press. She would also travel on some of the airship's later flights.

The airship left Friedrichshafen at 07:54 hours on 11 October and headed for the south of France to avoid bad weather reported over the Bay of Biscay. By early evening the craft had reached Barcelona, but more bad weather forced Eckener to continue flying south before heading back to the Azores via Madeira. Over Funchal, by which time the ship was travelling at 85mph, a bag of mail was dropped to the German consul.

On the morning of the 13th, when *Graf Zeppelin* was 1,100 miles (1,760km) east of Bermuda and 1,500 miles (2,400km) from Charleston, it ran into a squall. A sudden gust of wind damaged the port horizontal fin and speed was immediately reduced to 40mph (64kph). Eckener had uncharacteristically entered the squall at full power. He normally reduced speed in bad weather and as a result the ship pitched violently upwards with an inexperienced operator handling the elevators. Other airships had broken up under similar circumstances.

Eckener and his officers regained control of the ship, but not before a part of the fabric covering the port fin had been ripped away, threatening further damage. This could have made the ship uncontrollable. Eckener sent a repair team of four men – including his son, Knut – to make an in-flight repair. He also sent out a distress call, knowing that he was risking the reputation of his brand-new ship, and perhaps the entire Zeppelin enterprise. The distress signal was picked up by the press and newspapers around the world ran sensational stories about the looming destruction of *Graf Zeppelin*. The distress signal was received in Washington and warships were placed on standby.

The emergency repairs were successful, but the ship encountered a second squall. It emerged intact, but progress continued to be slow. Eckener had been expecting to reach Lakehurst on the 14th, but by early afternoon the ship was still near Bermuda. By mid-afternoon on the following day it was reported to be approaching the US coast, but the landing party at Lakehurst was concerned at the lack of wireless messages from the airship. It later transpired that the wireless operator had been kept busy sending out press reports.

Crossing the US coast, the *Graf Zeppelin* headed not for New Jersey but for Washington, Baltimore, Philadelphia and New York, so that the new airship could be shown off to the American public. That evening, though, *Graf Zeppelin* arrived safely at Lakehurst, having taken four days fifteen hours to travel from Germany. Eckener, his passengers and crew were greeted like heroes with a ticker tape parade along Broadway. President Coolidge told Eckener that his arrival was 'a symbol of the advance in air transportation which has been so ably furthered by your own efforts and those of your compatriots'.

The journal *Flight*, however, was not impressed. It acknowledged that the *Graf Zeppelin* had shown that the airship was not just a 'fair weather craft', but pointed out that its crossing had been no swifter than an ocean liner. Possibly the airship had been sent on its long journey 'somewhat prematurely and before a sufficient number of really exhaustive test flights had been made'. It also pointed out that nine years earlier the smaller *R34* had made the journey in a shorter time. 'The British airship,' *Flight* added, 'had neither the range nor the carrying capacity of the latest German vessel and thus her flight was in many ways more meritorious.'

But the *Graf Zeppelin's* return flight took seventy-one hours forty-nine minutes, just under three days and half the time taken by the swiftest ocean liners. After two weeks, during which more permanent repairs could be made to its damaged fin, the airship left Lakehurst. The homeward route was a more northerly one than taken on the outbound flight. On board were twenty-three passengers, including Mrs Clara Adams, who became the first woman passenger to buy a ticket to travel across the Atlantic. It also carried thirty-two mail bags.

Graf Zeppelin's next major flight, around the world, was probably its most famous. The driving force was newspaper magnate William Randolph Hearst, who paid $200,000 for exclusive US and British media rights to the story. On board were four of his journalists, including Grace Drummond-Hay and photographer and newsreel cameraman Robert Hartmann. The carriage of souvenir mail also helped offset the cost of the trip.

Hearst wanted the flight to begin and end in the USA, but the Germans disagreed. In the end there was a compromise: two flights in one with two starting points, Lakehurst and Friedrichshafen. The ship left New Jersey on 7 August 1929, reaching Friedrichshafen fifty-five hours minutes later to refuel before setting out across eastern Europe to the USSR and Siberia. On 19 August *Graf Zeppelin* arrived at Tokyo before making the first non-stop aerial crossing of the Pacific. It arrived at Los Angeles on 26 August and fifty-one hours fifty-seven minutes later reached Lakehurst. By 4 September *Graf Zeppelin* was back home in Friedrichshafen, having covered 20,650 miles (33,040km) in twenty-one days, five hours and thirty-one minutes to complete the fastest circumnavigation of the globe.

By the summer of 1931 *Graf Zeppelin* was ready to begin what would be the world's first scheduled, non-stop, intercontinental airline service. The South American market was considered ideal for *Graf Zeppelin* as Brazil and Argentina had substantial German communities with strong business and trade connections with the Fatherland. *Graf Zeppelin* promised to cut journey times from weeks to days and, consequently, the airship services were expected to be very popular.

They were. *Graf Zeppelin* crossed the South Atlantic eighteen times in 1932, and made a similar number of flights in 1933. By 1934, the Zeppelin company was advertising a regular service to South America, departing Germany almost every other Saturday to Brazil with connecting aircraft flights to Argentina. In 1935 and 1936 *Graf Zeppelin*'s schedule was virtually devoted to carrying passengers and mail between Germany and Brazil. There were round trips almost every fortnight from April to December. These operations represented the airship's principal function until they were withdrawn in May 1937. By that time *Graf Zeppelin* had made a total of sixty-four round trips to Brazil.

With the viability of long-range passenger operations by airship amply demonstrated, Eckener and the Zeppelin company wanted to build a fleet of similar vessels specifically designed for intercontinental commercial operations. This led to the design of *LZ-128*, which would have had capacity for 5,307,000 cu ft (150,294 cu m) of hydrogen. But the loss in October 1930 of the British airship *R101*, in which passengers and crew were killed by the hydrogen fire that followed a crash, convinced the Zeppelin company to switch to helium. Although heavier than hydrogen, it was less flammable.

The change to helium required a vessel able to carry 7 million cu ft of gas (198,240 cu m). The result was *LZ-129*, a ship considerably larger than *LZ-128*. In fact, *LZ-129*, to be named *Hindenberg*, and its sister ship, *LZ-130*, *Graf Zeppelin II*, remain the largest objects ever to fly, more than three times longer than an Airbus A380. The Nazi party was quick to spot the propaganda potential of this giant new airship. Construction had started in 1931, but progress was slow due to a shortage of funds following the worldwide depression.

At first the Nazi Party's assumption of power in January 1933 had little impact on the Zeppelin company. This was partly due to air minister Hermann Göring's dislike of lighter-than-air craft, but propaganda minister Joseph Goebbels saw the new airship as a showcase for Nazi Germany and accordingly offered 2 million marks towards *LZ-129*'s completion. Not to be outdone, Göring put up 9 million marks from his air ministry funds. There were, though, strings attached. In March 1935 the RLM, the German Air Ministry, split the Zeppelin company into two. The original *Luftschiffbau Zeppelin* would be responsible solely for [airship construction, while the newly created Deutsche Zeppelin-Reederei (DZR) – partly owned by the national airline Deutsche Luft Hansa – would operate them.

In a speech to mark DZR's formation, Göring expressed the hope that the *Hindenburg* would 'also fulfil its duty in furthering the cause of Germany'. In a clear indication of the Nazis' plans he declared: 'The airship does not have the exclusive purpose of flying across the Atlantic but also has a responsibility to act as the nation's representative.'

One result of DZR's creation was the end of Hugo Eckener's leadership. He had never been a Nazi supporter and his withholding of permission for Hitler to hold a rally in a Zeppelin hangar in 1931 had not been forgotten. Consequently, Eckener became little more than a figurehead in the company as the power passed to Ernst Lehmann, who was considered more sympathetic to the Nazis.

The party unashamedly exploited the huge and impressive airship. They frequently called on *Hindenburg* for propaganda flights, often in company with *Graf Zeppelin*. *Hindenburg* was a major presence at major public events including the 1936 Berlin Games and the Nuremberg Party rally. In fact, the first major trip after the completion of its flight test programme was a seventy-four-hour tour to support Hitler's remilitarisation of the Rhineland. The craft had made its first flight on 4 March 1936. It lasted three hours six minutes and on the 23rd the ship carried passengers for the first time when eighty reporters were treated to a short flight from Friedrichshafen to Lowenthal.

On 6 May 1936 *Hindenburg* began the service it was designed to undertake. By then, though, attitudes in the USA were changing. Two US Navy vessels had been lost and the American public now considered them dangerous. The USA had a monopoly on the non-flammable helium gas that the *Hindenburg* was designed to use, but the Helium Control Act of 1927 prohibited its export. Even more crucial was the use of the facilities at Lakehurst, New Jersey, the only terminal in the USA suitable for the *Hindenburg*. As it was a US Navy station, only the president could permit its use for commercial purposes. And despite his reduced status, there was only one man who could ask him for permission.

But for all his contacts in the USA it was only with great difficulty that Eckener was able to gain an audience with President Roosevelt. Eventually, in February 1936, he was received at the White House and granted the permission he sought. It seemed that the USA had not lost all interest in commercial airship operations.

The fifty-strong passenger list for the maiden trans-Atlantic voyage to the USA largely comprised celebrities, journalists and members of the Nazi elite, although there were a few fare-paying travellers with sufficiently deep pockets. During the crossing, Roman Catholic priest Fr Paul Schulte said mass, the first time this had ever been done in flight. The journey was also featured in a broadcast over the NBC radio network that included a recital played on *Hindenburg*'s specially made lightweight duralumin piano.

The crossing took two-and-a-half days, half the time normally taken by the *Queen Mary*. Even faster was the crossing made on 10 August, which took forty-three hours two minutes. But the saving in time was expensive: the one-way fare in 1936 was $400, rising to $450 the following year. This compared with the equivalent first-class fare on a ship of $157. *Hindenburg* also derived income from the large quantities of mail it carried, although its irregular flight dates during its maiden season, coupled with the high price of postage, meant that most of the letters carried in 1936 were special items designed especially for stamp collectors or souvenir hunters. Plans to introduce a more regular schedule in 1937, and possibly lower the cost of postage, left DZR optimistic that significant income could be earned by carrying business mail.

Indeed, eighteen return flights between Germany and the USA were scheduled for 1937 as sister ship *LZ-130* was nearing completion at Friedrichshafen. Over the winter *Hindenburg* underwent maintenance and renovation. As the airship was being operated with hydrogen rather than helium it had a greater lifting capacity and additional passenger cabins could be provided.

Although *Hindenburg* made six successful flights in 1937, the craft's first North American trip of the year was to be its last. The ship passed over New York and circled the Empire State building before heading for Lakehurst. Its arrival was delayed by stormy weather, but the ship was finally cleared to land when flame was seen sprouting from the rear. Within a frighteningly short time, fire had engulfed the whole ship. The outer covering was consumed rapidly, exposing the metal framework. What was left of the airship fell to lie on the earth like a giant smouldering skeleton.

Of the thirty-six passengers, thirteen were killed together with twenty-two of the sixty-one crew, and a civilian member of the ground handling team.

Ernst Lemann did not recover from the burns he received but Max Pruss survived. Subsequent investigation – Eckener was a member of the German delegation – revealed that the *Hindenburg* had been destroyed by escaping hydrogen ignited by a spark. There had been little evidence to go on but the possibility of sabotage was rejected.

Eckener believed that static electricity had ignited leaking gas. In view of the extensive precautions against fire on the *Hindenburg* it was an ironic end for the great silver fish. The era of transcontinental passenger Zeppelin travel, which had begun with such high hopes, was now over.

Graff Zeppelin II was completed but used only for training and propaganda fights. Just before the outbreak of the Second World War the ship made several flights around the British coast in an attempt to probe the nation's radar defences, but it failed to gain any significant information.

The Ultimate Airship

In its design and construction *Hindenburg* represented the ultimate evolution of the Zeppelin airship. Chief designer Ludwig Durr schemed a framework comprising triangular duralumin girders forming fifteen main rings, connecting thirty-six longitudinal girders with a triangular keel at the bottom of the hull and a cruciform tail. A corridor ran through the centre of the ship.

Although just 30ft (9m) longer than *Graf Zeppelin*, *Hindenburg* carried twice the volume of lifting gas due to its larger diameter. This also endowed it with greater structural strength against bending stresses than the *Graf Zeppelin* with its slimmer profile.

While previous Zeppelins had featured gas cells made of goldbeater's skin – the outer membrane of cattle intestines – *Hindenburg* used a new material created by sandwiching gelatine between sheets of cotton. Yet although the vessel was designed to use helium, the USA had a monopoly on the non-flammable gas and the Helium Control Act of 1927 prohibited its export.

Hindenburg's gas cells had fourteen manually controlled manoeuvring valves located just above the central walkway. They could be operated from the control car. To avoid damage to the cells or to the ship's framework, fourteen automatic valves released gas whenever cell pressure became too high.

The craft was propelled by Daimler-Benz diesel engines – safer than petrol – designed originally for high-speed motor torpedo boats as part of the Nazi rearmament programme. Each had a maximum power output of 1,320hp at 1,650rpm, although normal cruise setting, not usually adjusted during an ocean crossing, was 1,350rpm. At this speed an output of approximately 850hp was generated. The engines used compressed air starters and could be started, stopped and reversed in flight. Each drove a four-bladed, fixed-pitch, metal-sheathed wooden propeller of 19ft 6in (5.9m) in diameter using a 2:1 reduction gearing.

Unlike the *Graf Zeppelin*, the *Hindenburg*'s passenger accommodation was contained within the hull rather than in a gondola slung beneath it. The passenger space was spread over two superimposed decks, 'A' and 'B'. 'A' deck contained the 47ft-long (14.3m) dining room, which occupied the entire port side, as well as the lounge, writing room, port and starboard observation decks and twenty-five double berth cabins. 'B' deck contained the kitchen, passenger toilet and shower, officers' and crew's mess. Before the 1937 season, additional cabins were added to this deck.

Everything on board the ship was designed to be lightweight. This ranged from the aluminium dining room furniture – the chairs could be lifted with one finger – to the baby grand piano made of duralumin and covered in yellow pigskin. It was not carried on the ship's last voyage.

The berths were superimposed and could be folded against the wall during the day. An aluminium ladder provided access to the top bunk. There was a small fold-down desk, a white plastic wash basin and a small wardrobe covered with a curtain. The promenade decks featured seating and large windows that could be opened in flight.

Perhaps the most surprising feature was the smoking room. It was kept at a higher ambient pressure than the rest of the accommodation to avoid the risk of hydrogen leaking inside. The room and its associated bar were separated from the rest of the ship by a double-door airlock. This area was supervised by steward Max Schulze, who not only served cocktails but ensured the nobody left with lighted cigars, pipes or cigarettes.

Specifications

	Graf Zeppelin	*Hindenburg*
Crew	36	40
Passengers	20	50–72
Length	776ft (236.53m)	803ft 10in (245.3m)
Diameter	100ft (30.48m)	135ft (41.2m)
Volume	3,707,550 cu ft (105,000 cu m)	7,100,000 cu ft (200,000 cu m)
Useful lift	132,000lb (60,000kg)	22,046lb (10,000kg)
Power plant	five Maybach VL-2 twelve-cylinder engines each developing 550hp (410kW) for take-off	four Daimler-Benz DB 602 sixteen-cylinder diesels each developing 1,100hp (735kW)
Maximum Speed	80mph (128kph)	81mph (131kph)

Walter Wellman's airship America viewed from the RMS Trent during the final moments of his trans-Atlantic crossing attempt. The airship is seen here dragging its anchor in the water. The crew were eventually rescued by Trent in the Atlantic near Bermuda. (US Library of Congress; LOT10962)

Walter Wellman, seen here on the left, with an unidentified associate at an aviation meeting in October 1910. (US Library of Congress; LOT10962)

The Curtiss flying boat NC-4 which completed the first ever aerial crossing of the Atlantic in 1919. (US Library of Congress; LC-B2-4902-13)

NC-4 takes off from the Azores, bound for Lisbon, Portugal, during its trans-Atlantic crossing. (US Library of Congress; LC-F82-2960)

The captain of NC-4, Lieutenant Commander Albert Cushing Read. Later in 1919, upon returning to the US, Read predicted that, 'soon [it] will be possible to drive an airplane around the world at a height of 60,000 feet and 1,000 miles per hour'. (US Library of Congress; LC-F8-5343)

Two young RAF officers, Captain John Alcock and Lieutenant Arthur Whitten Brown, made the first non-stop trans-Atlantic flight on 14 June 1919, for which they shared a £10,000 prize awarded by the Daily Mail. (Historic Military Press)

Alcock and Brown taking off in their Vimy from Newfoundland, 14 June 1919. (Historic Military Press)

In the summer of 1979, two RAF McDonnell Douglas F-4 Phantoms were specially painted in 'Alcock and Brown' colours to commemorate the anniversary of the pair's trans-Atlantic flight. The paint work was undertaken by Warrant Officer John Cooper, Sergeant Ken Lillico and a volunteer team. The paint scheme was applied to XV424 and XV486, having been specially designed by aviation artist Wilf Hardy. The aircraft were finished in overall gloss light aircraft grey, with a swept red, white and blue fuselage stripe with a raked Union Jack on the fin, all of which was accompanied by commemorative inscriptions on the nose and underwing fuel tanks. (Historic Military Press)

The British dirigible R34 pictured just moments from the end of its epic trans-Atlantic flight approaching the landingsite at Mineola, Long Island, New York, on 6 July 1919. (US Library of Congress; LC-B2-6143-9)

The landing of R34 at Mineola, 6 July 1919. (US Library of Congress; LC-B2-4966-13)

The captain of R34, George Herbert 'Lucky Breeze' Scott pictured soon after the airship's arrival at Mineola. For his part in the record-breaking flight, Scott was made a CBE. (US Library of Congress; LC-B2-4971-15)

Charles Lindbergh prior to his record-breaking flight. (US Library of Congress; LC-B2-5311-6A)

Lindbergh with Spirit of St Louis. (US Library of Congress; LC-F8-4157)

Charles Lindbergh's Spirit of St Louis on display on a US Navy barge off Hain's Point, Washington DC, circa 1927. (Historic Military Press)

On the left in the back, Charles Lindbergh is driven up lower Broadway, New York, in an open car to celebrate his record-breaking flight. (US Library of Congress; 2001700260)

Two aviation pioneers meet. Orville Wright, on the left, with Colonel Charles Lindbergh, on the right, during the latter's visit to pay Wright a personal call at Wright Field, Dayton, Ohio, on 22 June 1927. In the centre is Major John F. Curry. (US Library of Congress; LC-W86-174)

General Francseco de Pinedo, on the left, pictured with the Italian Ambassador to the US at the time of his epic 'Four Continents' flight. (US Library of Congress; LC-F8-41376)

Dieudonne Costes (left) and Joseph le Brix (on the right) are greeted by President John Coolidge (second from left) at the White House, 9 February 1928. The French Ambassador to the US, Paul Claudel, is to the fourth individual in this picture. (US Library of Congress; LC-H2-B-2550)

Graf Zeppelin over the US capital. (US Library of Congress; LC-F81-43591)

Graf Zeppelin landing at Lakehurst, New Jersey, following its first trans-Atlantic crossing.

British Airways' Concorde G-BOAC in May 1986. Considered the flagship of BA's fleet because of its registration, G-BOAC first flew on 27 February 1975. It is preserved in Manchester, where it is on public display. (Eduard Marmet Collection)

A side view of a Lockheed SR-71, the fastest aircraft to ever cross the Atlantic, manoeuvring towards a KC-135 Stratotanker aircraft for inflight refuelling, January 1983. (USAF)

Norwegian's Boeing 787-9 Dreamliner G-CKHL which became the fastest subsonic aircraft to cross the Atlantic on 15 January 2018. At the time of the flight, G-CKHL carried tail fin artwork depicting Amy Johnson, the pioneering pilot who was the first female to fly solo from the UK to Australia in 1930. (Norwegian)

Captain Harold van Dam in the cockpit of Norwegian's Boeing 787-9 Dreamliner G-CKHL after the record-breaking flight on 15 January 2018. (Norwegian)

Chapter Eleven

Non-Stop to New York

The giant four-engine flying boat named *Lisbon Clipper* and operated by Atlantic Airways had cabins as roomy as an ocean liner, a luxurious dining room and an open-air promenade deck. It was an advanced specification by any standards, but in the 1930s it must have seemed like science fiction. It was.

Released in 1937, *Non-Stop New York* was a British made B-movie filmed largely at the British Gaumont studios at Shepherd's Bush in west London. Its key prop was a huge model flying boat used for both exterior and interior scenes. A smaller one was used for longer shots. The modelmakers' work was so realistic that one reviewer said the film's thrilling climax was less believable than the aircraft in which it was supposed to have taken place.

In those pre-war days, trans-Atlantic passenger services by flying boat – or any other aeroplane for that matter – happened only in fiction. Yet *Non-Stop New York* was not as far-fetched as the critics thought. In fact, it was remarkably prescient. While cinema audiences were watching the thrilling climax aboard the giant airliner, the Boeing company in Seattle was actually building just such a leviathan that would be carrying passengers across the Atlantic in two years' time.

Even though aircraft could exceed 400mph and fly higher than Mount Everest, the North Atlantic was still presenting insurmountable challenges in the 1930s. The ability to combine range with a commercially viable payload still seemed out of reach for aeroplanes. Only airships with lift provided by huge volumes of gas were capable of offering passenger air services over the Atlantic.

But by early 1937 only the *Graf Zeppelin* and the *Hindenburg* were actually doing so. In Britain and America such craft were now considered dangerous. The British rigid airship *R100* had, indeed, made a successful return trip from Cardington, Bedfordshire, to Montreal in July and August 1930 on what was intended to be a proving flight for regular scheduled passenger services. But the nation's airship programme was abandoned after the loss of the *R101* in October of the same year.

The dramatic loss of the *Hindenburg* meant the end of Germany's love affair with lighter-than-air craft, but as the passenger-carrying airship passed into history there were significant developments taking place in heavier-than-air technology. The all-metal Douglas DC-3 had brought new levels of passenger comfort and convenience to air travel. It could cruise at more than twice the *Hindenburg*'s speed and, although lacking the Zeppelin's long-distance load-carrying ability, could cross the USA eastbound in fifteen hours, albeit with three refuelling stops.

The logical next step for the airlines was the acquisition of aeroplanes with trans-Atlantic capability. There had been machines with intercontinental pretensions, but none could rival the Zeppelins. That included the giant Dornier Do X flying boat. Revealed in Germany in 1929, it was powered by twelve engines mounted above the wing in tandem pairs.

Perhaps the makers of *Non-Stop New York drew* inspiration from the Do X. Its luxurious passenger accommodation arranged on three decks certainly approached the standards of trans-Atlantic liners. On the main deck was a smoking room with its own bar, a dining room and seating for sixty-six passengers. The seats could be converted to sleeping berths for night flights. The cockpit and compartments for navigation, engine control and radio operation were on the upper deck. The lower deck held fuel tanks and nine watertight compartments, only seven of which were needed to provide full flotation.

Initially, the engines were barely able to lift such a huge machine, but a change of power plant enabled it to prove its load-carrying in October 1929 by lifting 169 people. A year later, in November 1930, it left on a demonstration flight to New York by way of West Africa, Brazil and the Caribbean. The journey was interrupted several times by mishaps and delays, so it was not until August 1931, almost nine months after leaving Friedrichshafen, that the aircraft reached its destination. While never a commercial success, the Dornier Do X was the largest heavier-than-air aircraft of its time. In fact, it was ahead of its time for it demonstrated the potential of intercontinental passenger operations, even if it was not itself a viable proposition.

More serious were the operations involving small floatplanes catapulted from ships to carry mail the last few hundred miles to their destinations. It had started in 1927 when Norddeutsche Lloyd (NDL) embarked a Junkers F.13 floatplane on its liner *Lutzow* to provide joy rides for passengers. The aircraft was water-launched and lifted on and off the ship by crane, but maintaining it proved difficult for the ship's crew. Nevertheless, it proved popular and sparked ideas about using ship-launched aircraft to carry mail across the Atlantic.

The Heinkel company designed a suitable catapult, which equipped the new NDL liners *Bremen* and *Europa*. Germany's national airline, Deutsche

Luft Hansa (DLH), agreed to support the venture and when *Bremen* departed on its maiden Atlantic voyage in 1929 it carried a single-engined Heinkel He 12 floatplane, registered D-1717. The aircraft was a conventional strut-braced monoplane with tandem open cockpits for the pilot and radio operator. The mail was carried in a compartment behind them.

When the *Bremen* was 248 miles (397km) from New York, the aircraft, carrying 660lb (300kg) of mail and piloted by DLH's Captain Jobst von Strudnitz, was catapulted from the ship. It landed in New York Harbour two hours thirty minutes later, to be greeted by the city's mayor. The process was repeated on the return voyage when the liner was near Cherbourg. The aircraft reached the port of Bremerhaven 600 miles (960km) away twenty-four hours ahead of the liner. The mail was immediately flown on to Berlin in a waiting DLH aircraft.

Eight further flights were made during the year. By March 1930 Martin Wronsky, DLH's general manager, was able to announce in a lecture to the Royal Aeronautical Society in London that the travelling time for mail between New York and Berlin had been almost halved to thirty-six hours.

Over the next few years improved catapults enabled bigger aircraft to be used. Initially this meant Heinkel He 58s, but by 1934 a pair of the more advanced single-engined enclosed cockpit six-passenger Junkers 46s (D-2244 and D-2271) were embarked on *Bremen* and *Europa*.

These services were not confined to the North Atlantic. Between February 1934 and August 1939 DLH operated a regular airmail service between Natal, Brazil, and Bathurst, Gambia, continuing via the Canary Islands and Spain to Stuttgart. At first it used Dornier Wal flying boats on the oceanic stage with a refuelling stop in mid-ocean. This entailed landing on the open sea, near a converted merchant ship that was equipped with a 'towed sail', on to which the aircraft taxied before being winched aboard. After refuelling it was launched by catapult to continue its journey.

From September 1934 a second ship was available to provide a support vessel at each end of the oceanic stage. Heinkel He 70s, which, with a top speed of 200mph (320kph) represented the peak of contemporary high-speed airliner development, were used for the leg between Berlin and Seville. From there the mail went by Junkers Ju 52, with flying boats making the actual crossing.

When the mail from Europe arrived in West Africa from the Canary Islands by Wal the support ship carried the aircraft towards South America and launched it by catapult after thirty hours. On the return journey the Wal flew from Natal to Fernando de Noronha before being craned aboard the ship at night to be catapulted off for West Africa in the morning. From April 1935 the Wals were launched offshore, and flew the entire distance across the ocean, cutting the time taken for mail to reach Brazil from Germany from four days to three.

The ships used on the South Atlantic service were the SS *Westfalen* and the MS *Schwabenland*. A purpose-built seaplane tender, the MS *Ostmark*, went into service in 1936 and continued until 1938.

Aircraft of more recent design were introduced from 1937. The Dornier Do 18s, which supplanted the Wals together with the bigger four-engined Blohm and Voss Ha 139s, used diesel power. Heavy oil engines were comparatively new to aircraft, the first entering service with DLH in 1931. But after overcoming initial difficulties, the Jumo 205 engines were considered well-suited to long-distance operations, offering lower fuel consumption, use of cheaper fuel and reduced fire risk. Diesel-powered Do 18s made six North Atlantic crossings in 1936, while by mid-1938 the Ha 139 floatplanes, which were introduced in 1937, had completed fourteen.

They were operated on the South Atlantic between Bathurst and Natal. Ship-launched, they could carry 1,100lb (500kg) of mail over 3,125 miles (5,000km). With the outbreak of war, the aircraft were transferred to the *Luftwaffe*, but were not well suited to military duties.

From 1927 the French Compagnie General Aeropostale had operated an airmail service on the South Atlantic route between France and Brazil. Initially the mail was flown only as far as Dakar in Senegal, West Africa, before being loaded aboard converted destroyers for the rest of the journey. Air France acquired Aeropostale and continued to operate the service, which took a week to complete.

By the late 1930s it seemed clear that water-borne aircraft represented the best option for trans-Atlantic airliners. In the US, as in Germany and France, attention was initially focused on a faster mail service between America and Europe. In 1931 US postal official W. Irving Glover wrote an article for *Popular Mechanics* in which he called for the construction of mid-ocean 'seadromes'.

Meanwhile, the key challenge, apart from the distances involved, continued to be the weather. The route to America via the Azores and Bermuda was generally good all the year round with winds varying between light north-easterly and light westerly. By contrast, the northern route between Newfoundland and Ireland was afflicted with fog, frequent storms and heavy cloud. There was also the constant risk of icing. One thing, the journal *Flight* noted in December 1936, was certain: 'The whole success of the Atlantic project depends on the meteorologist and meteorological stations have been installed in Newfoundland for the purpose.'

It was at about this time that Juan Trippe, visionary president of Pan American World Airways, began seriously to ponder the feasibility of trans-Atlantic passenger services. In Britain the managers of Imperial Airways were thinking along similar lines. In fact, the state-owned airline was acquiring a fleet of modern flying boats. The four-engined Short C-Class Empire boats were primarily intended for operation on routes to Africa and Asia. The aircraft lacked the range of the Sikorsky Clippers Pan American was using on its Pacific routes, but two C-Class boats, *Caledonia* and *Cambria,* were lightened and equipped with long-range tanks to increase range to 3,300 miles (5,280km).

In late 1936 *Flight* was relishing the launch of services between Bermuda and New York, which it described as 'one of the most magnificent air transport experiments that have ever been attempted'. It added: 'Only a very phlegmatic person could fail to be profoundly stirred by plans which are, nevertheless, being carried out with matter-of-fact precision and with an entire lack of unnecessary heroics.'

The journal was anticipating the start of flying boat services by Imperial Airways and Pan American between Bermuda and New York. It was only seven years earlier that Bermuda had first been linked to the USA by air. On 1 April 1930 three Americans, Lewis Alonzo Yancey, William H. Alexander and Zeh 'Jack' Bouck, flew the 700 miles (1,200km) from Long Island to Bermuda to claim the $25,000 prize put up by the island's government to stimulate tourism. The only navigational aids carried by their single-engined Stinson SM-1FS Detroiter floatplane were a compass and a single US Navy survey map. When the inevitable happened and the intrepid trio lost their way, they had to spend a night on the water before resuming their journey the next morning.

On 16 June 1937 Imperial Airways' C-Class flying boat *Cavalier* (G-ADUU), which had previously been shipped to the island and reassembled there, took off from the uncompleted marine air terminal at Darrell's Island with Captain Neville Cumming in command. At the same time Pan American's Sikorsky S-42, (NC 16735, christened *Bermuda Clipper* by Mrs Juan Trippe) left Port Washington, New York, for Bermuda with Captain R.O.D. Sullivan in command. On 16 November, Pan American began flights to Bermuda from Baltimore, its winter base. *Cavalier* continued operations until 1939 when it was lost in mid-Atlantic while en route from the USA to Bermuda. The aircraft was forced down by icing and three of the thirteen on board were killed.

But both airlines had more ambitious plans. In 1935, anticipating the start of trans-Atlantic flying boat services, the governments of the US, Britain, Canada and the Irish Free State had reached an agreement for the Irish port of Foynes, the most westerly in the country, to act as the European terminal for trans-Atlantic flying boat operations.

On 5 July 1937 Captain A.S. Wilcockson flew *Caledonia* from Foynes to Botwood, Newfoundland, which was to provide the terminal at the other end of the service, while Captain Harold Gray piloted a Pan American Sikorsky S-42 in the opposite direction. *Caledonia* covered the 1,900 miles (3,040km) in fifteen hours three minutes at an average speed of 132mph (211kph). 'It was a great trip,' Wilcockson commented. 'It shows that a regular transatlantic mail and passenger service is quite feasible.' In August, another Imperial Airways C-Class captain, Griffith 'Taffy' Powell, completed the crossing from Foynes to Botwood in *Cambria* in fourteen hours twenty-four minutes. Indeed, both operations had been successful and were followed by a series of proving flights in a variety of weather conditions. On 16 August 1937 Pan American began a series of flights to and from Bermuda with stops at the Azores, Lisbon, Marseilles and Southampton.

The following month, *Cambria* took just ten hours thirty-three minutes to make the return crossing. These feats earned Powell recognition in one of a series of Kings of Speed cigarette cards issued by Churchman's. 'As a matter of fact,' Powell wrote many years later, 'it was a record that stood for many years.'

But it was still a fact that, even with enlarged fuel tanks, the Short C-Class boats could have range or they could have passenger capacity, but not both. One solution, proposed by Major Robert H. Mayo, Imperial Airways' technical general manager, was to mount a small, long-range seaplane on top of a larger carrier aircraft and use their combined power to bring the smaller aircraft to operational height. At that point the two would separate, the carrier aircraft returning to base while the other flew on to its destination.

The Short-Mayo Composite project, co-designed by Mayo and Short's chief designer Arthur Gouge, comprised the Short S.21 *Maia* (G-ADHK), a modified C-Class equipped with a pylon on the top of its fuselage, and the smaller S.20 *Mercury* floatplane (G-ADHJ) powered by four Napier Rapier sixteen-cylinder H-formation air-cooled engines. The aerofoil design of both aircraft was such that *Mercury*'s wings were carrying the major part of the air load at the speed and height selected for separation. Safety locks prevented separation until this speed had been reached and both pilots had simultaneously operated the separation controls. The first separation was made over the Thames estuary on 6 February 1938. It was a complete success. *Mercury* leapt upwards as planned while *Maia*, deprived of the lift provided by the smaller aircraft, went into a shallow dive.

In the trans-Atlantic crossing, which followed on 21 July, *Mercury* separated from *Maia* to complete what would become the first commercial non-stop east–west trans-Atlantic flight by aeroplane. *Mercury* took twenty hours twenty-one minutes to reach Boucherville, Quebec, from Foynes. When it passed Newfoundland's Cape Bauld thirteen hours twenty-nine minutes after separating from *Maia, Mercury* had completed the fastest east–west crossing yet made, beating the previous best of fourteen hours twenty-four minutes set by *Cambria* the previous August.

Mercury was commanded by Captain Donald Bennett, a highly regarded Imperial Airways pilot and navigator who would figure in the trans-Atlantic story in future. He helped establish the wartime ferry service and later became managing director of British South American Airways. As the leader of RAF bomber Command's wartime Pathfinder Force, the Australian-born Bennett had been the service's youngest air-vice marshal.

According to aviation historian Geoffrey Norris, the Short-Mayo Composite was so successful that it would have led to a landplane version had not the war intervened. Even so, Norris has called it, 'one of the most remarkable aerial experiments of all time'.

Another technology developed to facilitate trans-Atlantic commercial operations was aerial refuelling. Using dedicated tanker aircraft, in-flight refuelling would become a routine feature of military operations, but in the mid-1930s it was certainly something new. Sir Alan Cobham developed the grappled-line, looped-hose system for long-range trans-oceanic commercial operations and it was publicly demonstrated for the first time in 1935.

The receiving aircraft trailed a steel cable, which was then grappled by a line shot from the tanker. The line was then drawn back into the tanker, where the receiving aircraft's cable was connected to the refuelling hose. The recipient crew could then haul back the cable, bringing the hose to it. Once the hose was connected, the tanker climbed above the receiving aircraft to allow the fuel to flow to it by gravity.

Cobham had founded Flight Refuelling Ltd in 1934 and by 1938 had demonstrated the ability to refuel the Short C-Class boat *Cambria* from an Armstrong Whitworth AW.23 tanker. From 1939, converted Handley Page Harrow bombers were used to test the feasibility of refuelling the flying boats for regular trans-Atlantic crossings. Between 5 August and 1 October 1939, C-Class boats made sixteen crossings, with FRL's aerial refuelling system used on all but one. Further trials were suspended on the outbreak of the Second World War.

Meanwhile, an enlarged C-Class was emerging from Short's Rochester factory. The S.26 G-Class, powered by four 1,400hp (1,044kW) Bristol Hercules sleeve valve radial engines, was designed to cross the Atlantic without refuelling. Three were built and the first, G-AFCI, *Golden Hind*, made its maiden flight on 21 July 1939 with Shorts' chief test pilot John Lankester Parker at the controls. Although two aircraft were handed over to Imperial Airways for crew training, all three were impressed into RAF service, together with their crews, on the outbreak of war.

Meanwhile, in July 1936 Pan American ordered six Boeing 314 long-range flying boats with an option for six more. Capable of crossing the Atlantic non-stop, they were built for one-class luxury travel and would cruise at 188mph (303kph). They had a lounge and dining area, and chefs recruited from four-star hotels prepared five and six-course meals served by white-coated stewards. There were separate dressing rooms for men and women.

On 11 April 1939, *Yankee Clipper,* registered NC18603 and commanded by Captain Harold Gray, landed at Foynes to complete the first trans-Atlantic proving flight. The first mail flight followed on 28 June and two weeks later, on 9 July, *Yankee Clipper* arrived at Foynes at the end of the first commercial passenger service on the direct route between the USA and Europe. The modern streamlined shape of the giant flying boat seemed to herald a new era in air travel. Services continued until the war, when the fleet was pressed into military service to ferry personnel and equipment between the US and European and Pacific theatres of war.

Imperial Airways was replaced by the newly created British Overseas Airways Corporation (BOAC) on 1 November 1939. The trans-Atlantic link was resumed the following August when the C-Class flying boats *Clyde* (G-AFCX) and *Clare* (G-AFCZ), their silver pre-war finish now covered by camouflage paint, made six return crossings carrying passengers. On one of these flights, four of *Clare's* six passengers were the first American civilian pilots to be recruited for ferrying military aircraft to Britain. From 1941 BOAC's newly acquired Clippers operated trans-Atlantic flights with the C-Class boats *Champion* (G-AFCT), *Cathay* (G-AFKZ) and *Clare* maintaining the link between Foynes and Poole.

BOAC's Boeing 314s were known as *Berwick* (G-AGCA), *Bangor* (G-AGCB) and *Bristol* (G-AGBZ). The latter made the fleet's inaugural flight on 22 May, commanded by Captain J.C. Kelly Rogers, on the Foynes–Lisbon–Bathurst–Lagos route. For the rest of the war the three boats operated a Foynes–Lagos–Baltimore schedule, but as they required regular maintenance in Baltimore, trans-Atlantic passenger flights via Botwood were slotted into the schedule. In winter, when Botwood was iced up, they flew from Bathurst to Baltimore via Belem, Brazil, Trinidad and Bermuda.

BOAC and Pan American were not the only airlines operating trans-Atlantic services during this period. Between April 1942 and 1945 American Export Airlines made 405 crossings between Foynes and New York on behalf of the US Navy using Vought-Sikorsky VS-44A flying boats. The aircraft could carry sixteen passengers, cruised at 175mph and had a range of 4,545 miles (7,272km). They received the USN designation JR2S-1, but flew in civilian markings.

On one occasion American Export's chief pilot, Charles Blair, decided there was no need for a refuelling stop in Newfoundland and continued to New York, where the aircraft, *Excalibur 1* (NC41880), landed twenty-five hours forty minutes after leaving Foynes. It was the first non-stop commercial flight from Europe to a US city. On board was Admiral Sir Andrew Cunningham, commander-in-chief of the British Mediterranean Fleet, who described the trip as a 'remarkable voyage'. After the war, Blair made several notable long-distance flights in a converted Mustang fighter christened *Excalibur III*. He later married actress Maureen O'Hara, whom he had met during one of his 1,575 trans-Atlantic crossings.

By later summer 1945, with the war almost over, Pan American was able to demonstrate just how routine the trans-Atlantic operation had become. On 18 August *Atlantic Clipper* (NC18604) and *Dixie Clipper* (NC18605) both arrived at Foynes from New York, returning to the USA that night. Between them they carried a total of 101 passengers.

The Boeing 314

In 1935 Pan American boss Juan Trippe issued a specification calling for a flying boat capable of crossing the Atlantic. Although Boeing initially lacked interest in the project, it submitted what was to be the winning design.

Its Model 314 drew much inspiration from the company's one-off XB-15 bomber, marrying its wings and engine nacelles to the flying boat's towering, whale-like body. However, the use of new Wright 1,500hp Double Cyclone engines eliminated the power deficiency that had handicapped the XB-15. These double-row fourteen-cylinder radials were the first engines to use 100-octane fuel.

It was the biggest aircraft Boeing had yet built or, indeed, would build until the eight-jet B-52 bomber of 1952. Although it is possible to see these flying boats as the 'jumbo-jets' of their day, they provided luxury travel at fares that would, in equivalent terms, be double those charged for the Concorde supersonic airliner half a century later. On the Clipper's first passenger flight twenty-two passengers paid $375 for the one-way Atlantic crossing.

There was a six-man flight deck for the aircraft commander, two pilots, flight engineer, navigator and radio operator. There were four flight attendants, all male. Up to seventy-four day passengers could be carried seated or thirty-six in sleeper accommodation in a hull big enough to be divided into four separate compartments. Passengers had the benefit of large windows and enjoyed the comforts of dressing rooms, a dining salon that could be turned into a lounge and a bridal suite in the tail.

In place of wing tip floats to provide stability on the water, the 314s featured hull-mounted sponsons. The wing was thick enough to allow in-flight access to the engines through a walkway. Fully-feathering propellers made it possible for a mechanic to perform repairs in flight. Between June 1939 and June 1941 this facility was used on 431 occasions.

As the 314's single fin and rudder was found to provide insufficient directional control it was replaced by two and finally three. Aircraft with this modification were designated 314A and six were delivered to Pan American. Five of the original six were brought up to this standard.

Although it was originally scheduled before Christmas 1937, the prototype did not make its maiden flight until the following June. Delivery was delayed until January 1939. The flagship, *Yankee Clipper*, was christened by the first lady, Eleanor Roosevelt, in a ceremony at the Naval Air Station, Anacostia. She cracked a gold-trimmed bottle containing seawater rather than the customary champagne over the flying boat's blunt nose.

Despite the delayed service entry, the start of commercial Clipper operations represented a notable success for Boeing and for Pan American. Rival airlines could only marvel at the accomplishment. Few other contemporary aircraft could match the 314's range and load-carrying ability.

President Franklin D. Roosevelt travelled by Boeing Clipper to meet Prime Minister Winston Churchill at the Casablanca conference in 1943. On the way home, Roosevelt celebrated his birthday with a cake specially baked in the flying boat's galley. Churchill was also a Clipper passenger, at one point taking the controls of BOAC's *Berwick*. It made, he wrote later, 'a most favourable impression on me'.

Pan American's Clippers made around 5,000 ocean crossings and flew more than 12.5 million miles (20 million km), logging more than 18,000 flying hours each. During the war they carried more than 84,000 passengers, most of whom were on journeys of importance to the war effort.

Boeing 314A Specification

Accommodation	flight crew of six, four cabin stewards and 74 passengers (day, 36 night)
Length	106ft (32.33m)
Wingspan	152ft (46.36m)
Height	20ft 4.5in (6.22m)
Loaded weight	84,000lb (38,000kg)
Power plant	four Wright R-2600-3 radials each developing 1,600hp (1,200kW)
Speed	210mph (340kph) maximum, 188mph (302kph) cruise at 11,000ft (3,400m)
Range	3,685 miles (5,896km)
Service ceiling	19,600ft (5,980m)

Chapter Twelve

Atlantic Bridge

The group of men who arrived dishevelled and unshaven at a Belfast Hotel one morning in November 1940 must have seemed a distinctly odd bunch.

They spoke in a variety of unfamiliar accents, but even more unusual for wartime Britain was their attire. Some wore ten-gallon cowboy hats and high-heeled Texan boots, while others were in Canadian hooded parkas. One of them looked more like a bank manager in his black homburg hat with long dark overcoat and carrying a briefcase. No wonder they attracted so many curious glances in the hotel lobby.

The oddest thing about them, though, was that they claimed to have just flown in from England. It is not clear if the hotel receptionist was taken in by a piece of deception intended to preserve the security of an operation that could materially alter the course of the war, but the truth was they had just flown the Atlantic.

At a time when an oceanic crossing by air was still an unusual and, indeed, hazardous undertaking, these aviators had left Newfoundland the previous evening in seven brand-new Lockheed Hudson patrol bombers. They had arrived at Belfast's Aldergrove airport earlier that morning.

Even though the aircraft were in camouflage paint and displayed RAF roundels, most of their crews were civilians. Some were foreigners. There were nine Americans, six Britons, six Canadians and one Australian. That same evening, however, they were taken to Belfast Harbour and bundled aboard a ship bound for Canada.

The haste was vital. They were needed on the other side of the Atlantic as quickly as possible because many more American-built aircraft were waiting to be ferried to Britain to help the war effort. The need for these warplanes to supplement domestic production in the struggle against the Nazis was urgent.

To some extent this had been identified two years earlier in 1938 when a British purchasing commission led by senior civil servant Henry (later Sir Henry) Self was despatched to the USA to find out what was available. Top of the commission's shopping list was a light bomber and coastal reconnaissance aircraft capable of supplementing RAF Coastal Command's Avro Anson.

'Faithful Annie' was already approaching obsolescence and Lockheed's converted Model 18 transport seemed suitable as a replacement, especially as it could fly 60mph (96kph) faster than the Anson.

The subsequent order for 250 Hudsons for the RAF represented the largest single one so far gained by the youthful and energetic Lockheed company. But US neutrality laws prohibited war materials to be transported directly from American ports to a belligerent. The Hudsons had therefore to be flown by civilian Lockheed employees from the factory in Burbank, California, to the town of Pembina, North Dakota, 2 miles (3km) from Emerson, Manitoba. Today, Pembina is one of the busiest crossing points on the US–Canadian border, but in 1939 and 1940 it was the place from which lorries, or more usually horses, towed US-built aircraft bound for Britain across the border. Once on the Canadian side they were loaded on to trucks and driven to St Hubert airport, Montreal. From there they were transported to Halifax, Nova Scotia, to be prepared for the journey by ship to the UK. This meant they were either partially disassembled for carriage in the hold or simply protected against the sea and the weather as deck cargo. Often, the ships had to await the assembly of a protective convoy before sailing.

When they finally arrived in the UK the aircraft were transported to RAF Speke near Liverpool, where they were assembled and test flown. The whole process could take three months at a time when the Hudsons were desperately needed by RAF Coastal Command. The irony of the situation was that aircraft required to protect Britain's shipping convoys were themselves risking loss to prowling U-boats. They were also taking up valuable shipping space.

Canadian press baron Lord Beaverbrook, now Britain's Minister for Aircraft Production, was growing impatient with the delay. Why, he asked, could the aircraft not be flown over to Britain to speed up the process? In response, Air Ministry officials and senior RAF officers argued that, twenty years after Alcock and Brown's flight, there had been fewer than 100 successful North Atlantic crossings with fifty failed attempts, leading to the loss of sixteen aircraft and forty crew members. Such flights were, therefore, only to be undertaken by the most experienced airmen and certainly not in the winter.

However, George Woods Humphery, former managing director of Imperial Airways, advised Beaverbrook that it would, in fact, be feasible to fly the Hudsons across the Atlantic. Recently deposed following the establishment of British Overseas Airways Corporation and now living in the USA, Woods Humphery agreed to take on the task of supervising the operation, provided he had the support of experienced Imperial Airways captains. But he was careful to add that an existing organisation would be needed to provide administrative support.

Sir Edward Beatty, president of Canadian Pacific Railways, was an old acquaintance of Beaverbrook as well as Woods Humphery. A call from

Beaverbrook's office settled the matter. Since 1939 Beatty had been the Canadian representative of the British Ministry of War Transport charged with moving military supplies to the various theatres of war. Under his direction, a collection of bush-flying outfits acquired by Canadian Pacific started organising the ferrying of bombers to Britain. It was known as Canadian Pacific Air Services.

By August the captains recommended by Woods Humphery had arrived in Canada and established an office at Montreal's Wood Street Station. They included captains A.S. Wilcockson, R.H. Page, I.G. Ross and Donald Bennett, the last-named being the man who had commanded the record-breaking flight of the seaplane *Mercury*. All had navigating as well as piloting qualifications combined with experience of flying Imperial Airways' C-Class flying boats.

According to Canadian historian Dr Carl Christie, Bennett was 'probably the single most important individual in getting the Atlantic ferrying scheme launched successfully'. He and Page visited the Lockheed factory to satisfy themselves the Hudsons were capable of making the Atlantic crossing. They reported that the first batch would be ready by September.

But neither the RAF nor the Royal Canadian Air Force (RCAF) could spare crews to fly them. This meant that civilian volunteers would have to be recruited, most from the neutral USA. This was a sensitive matter and it was essential to avoid word reaching the Germans, who would, no doubt, have made every effort to intercept the aircraft before they reached the UK.

At the same time, an organisation established by a First World War American pilot called Clayton Smith and Canadian businessman Homer Smith was quietly recruiting US volunteers for the RCAF. Those with suitable qualifications were diverted to ferry duties following stringent scrutiny by Wilcockson and Bennett. By the end of November 1940, sixty-two had been identified as suitable for further training, which was provided at St Hubert airport south-east of Montreal, now the operation's headquarters.

Bennett drove them hard. 'We realised the enormous knowledge of aviation this man had,' one crew member recalled. Indeed, Bennett had devised a navigational procedure that would enable the Hudsons to maintain a set course and allow for frequent changes in wind strength and direction.

The aircraft were fitted with additional fuel tanks in the fuselage to enable them to make the crossing, a process that increased the dangers of an already hazardous undertaking. The Air Ministry in London had privately acknowledged that if only half the aircraft reached the UK the risk would be worthwhile. 'A willingness to accept such a high loss rate must surely speak volumes about the precarious situation in which the UK found itself,' says Christie.

Between 29 October and 9 November seven newly built Hudsons were flown from Montreal to the RCAF airfield at Gander, Newfoundland. Just over 200 miles (320km) from St John's, it had been built jointly with the British. By the autumn of 1940 it was about to become one of the world's busiest airports.

It was decided that, because of a shortage of suitably qualified navigators, the first batch of Hudsons would cross the Atlantic in loose formation. Bennett, who was supervising the operation, would fly in the leading aircraft, which would carry two radio operators, one British and one Canadian. The other six Hudsons also had a crew of three: pilot, co-pilot and radio operator. The pilots and co-pilots were British or American, but many of the radio operators had not flown before and none had made a long-distance flight.

The Hudsons left a snowy Gander on the evening of 10 November 1940 following a favourable weather forecast from Canadian meteorologist P.D. McTaggart-Cowan, also known to the crews as 'McFog'. As the aircraft took off the sound of their engines drowned out the Canadian military band playing *There'll Always be an England*. 'I remember getting quite a thrill from it,' one Hudson crew member would recall.

Supervising the operation for the RCAF was Squadron Leader Griffith 'Taffy' Powell. His main emotion was relief, as he later recalled: 'I was so immensely relieved to get 14 engines running after the endless difficulties with flat batteries, changing spark plugs and the constant cold wind.'

Over the ocean the night was dark and the weather decidedly mixed. The formation managed to stay together until they reached what Bennett described in his subsequent report as 'a very virile warm front'. The formation was forced to disperse at 18,000ft (5,490m) to avoid the risk of collision. Bennett climbed to 20,000ft (6,100m) and encountered heavy snow and turbulence. Although the aircraft had to continue the journey on their own, Bennett reported, 'All arrived safely and without incident.'

The crew of Hudson T9468 might not have agreed with this assessment. Things went well enough for the first hour, but then Captain Ralph Adams discovered a leaking oil tank. He monitored it closely and decided it was not bad enough to warrant turning back. But then there was a short circuit in the radio transmitter's aerial box, causing radio officer Curly Tripp to switch it off smartly to avoid the risk of fire.

Then the needle on the radio compass broke. 'I really felt up the creek without a paddle,' Tripp said later. To add to the discomfort of flying at 18,000ft, the Hudson's crew were breathing oxygen, sucking it through a rubber tube. Tripp managed to get the radio transmitter repaired and was able to contact Bennett before radio silence had to be imposed.

At 0800 hours GMT the co-pilot reported sighting land. After making a circuit of Aldergrove, T9468 landed fifty minutes later. 'We were third in,' Tripp noted. While the crews enjoyed breakfast in the officers' mess, Bennett continued to peer anxiously out of the window until all the aircraft were down safely. 'I have often wondered,' Tripp mused later, 'what his thoughts were as that seventh ship rolled to a stop. He had spent months of figuring and hours of hard work and finally himself led that first group to pioneer and prove it was possible to fly the North Atlantic in the winter.'

The second group of seven Hudsons left on 26 November. Their departure was twice postponed because of bad weather. The outside temperature at Gander had fallen to minus 9°C, but the aircraft were finally dug out of the snow, cleared of ice and set off led by Captain Humphrey Page. Six arrived in Northern Ireland the next day and one in England.

According to the official history, written anonymously for the Ministry of Information by best-selling author John Pudney, 'Trans-Atlantic ferrying was established beyond doubt. These airmen, backed by Canada's ground organisation, had overcome even the notorious weather conditions which hitherto had been regarded as making winter crossings impracticable. A great flow of traffic across the [Atlantic] Bridge could now be planned.'

Indeed, another seven Hudsons made it across on 17/18 December led by Captain Gordon Store. And a week later, on Christmas Day, the fourth group left Montreal. One of the aircraft, piloted by Bennett and christened *Spirit of Lockheed Vega*, was a gift to the people of Britain from the workers at the Burbank factory. This time, of the seven Hudsons, three failed to complete the journey. One crashed on take-off, without fatalities, but it blocked the runway for the second aircraft. A third had to turn back with engine trouble. Even so, by the end of 1940 twenty-five much-needed aircraft had been delivered safely in a fraction of the time it would have taken by sea.

With an increase in the supply of freshly trained navigators, formation flights were abandoned and aircraft were despatched individually. Prestwick on Scotland's Ayrshire coast had been chosen as the ferry operation's UK terminal due to its good weather record. The nerve centre of the operation was established in Redbrae House, which in peacetime would accommodate the first Scottish air traffic control centre.

In January 1941 another route was opened when the first of a batch of seven Consolidated PBY long-range patrol flying boats, known to the RAF as Catalinas, left Bermuda for the UK. For one crew it was to be a particularly harrowing trip. Six hours out, the automatic pilot suddenly jammed and the starboard aileron went to full down. The aircraft fell into a spiral dive. As it plunged towards the sea the crew desperately tried to lighten the aircraft by throwing out items of loose equipment. The Catalina finally recovered when it was just a few hundred feet above the ocean.

When daylight came the crew could see that both ailerons had been torn off. The aircraft was flying in the wrong direction, but gradually, with gentle use of the rudder, the pilots worked together to return it to the correct course. After an epic twenty-eight hours fifty minutes in the air the Catalina landed at Milford Haven. Taffy Powell, who was now in charge at Bermuda, later recalled the crews' fear of interception by the *Luftwaffe*'s long-range Focke-Wulf Fw 200 Condors as the lumbering flying boats neared the end of their journeys.

Also clear was the paramount importance of good weather forecasting and particularly the likelihood of icing. From 1937 the British Air Ministry and the Canadian Department of Transport had moved to create a meteorological organisation with forecasting centres established at Gander, Botwood and Montreal. Foynes in neutral Ireland was withdrawn on the outbreak of war. This organisation, together with an associated signals and control procedure, was to be the subject of an international agreement known as the Transport Air Services Safety Organisation, which was later modified to include the ferry operation.

Early in 1941 there were major personnel changes at the Canadian end of the operation. Woods Humphery had quit and C.H. 'Punch' Dickens, a well-known former bush pilot, had taken control of the Canadian Pacific Air Services Department. He was now planning for a major expansion of the infrastructure as traffic grew. Top of his list was a new airport for Montreal. Accordingly, the government was approached with a request to build one near Lake St Louis, 10 miles (16km) from the city. It would be called Dorval.

At the same time some of Canada's most prominent businessmen were being recruited to take key positions with the organisation at a nominal salary. They became known as the 'dollar-a-day-men'. One of them, Harold Long, was now in charge of administration at Montreal. But there were still some anomalies. Wilcockson was chief of operations, but Bennett was styled flying superintendent; there did not seem to be much difference between the two. The abrasive Australian had earned himself a reputation for getting things done, but his management style and that of Long were poles apart.

As the organisation grew it acquired a new name: the Atlantic Ferry Organisation, or ATFERO for short. But it was to be short-lived. In May 1941 it came under the control of the Ministry of Aircraft Production and Dickens is credited with helping to smooth the transition. In the USA the passage of the Lend-Lease Act was bringing even more changes. General Arnold, chief of the US Army Air Corps, proposed that American military pilots should take the aircraft to Montreal or another despatch point, releasing civilian pilots to fly the Atlantic and allowing British crews to be returned to operational duties.

This was subsequently agreed between President Franklin D. Roosevelt and Prime Minister Winston Churchill, but the president's stipulation that the aircraft had to be handed over to a military command rather than to a civilian organisation spelled the end of Canadian Pacific's involvement. On 28 May Roosevelt directed the Secretary of War to take full responsibility.

The British responded by placing the ferry operation under military control, too. In July 1941 ATFERO was subsumed into RAF Ferry Command. Air Chief Marshal Sir Frederick Bowhill, former chief of Coastal Command, was now in charge with his headquarters at Montreal. In reality Bowhill's command still included civilians both in the air and on the ground. At midnight on 14 July when the Canadian Pacific Railways contract officially terminated, ATFERO's establishment included 207 civilians, 118 RAF personnel and 18 from the RCAF. Taffy Powell was appointed Bowhill's senior air officer, a job he held until the end of the war in Europe.

Meanwhile, a further significant change had been made to the way the ferry operation was conducted. As it could take crews between ten and fourteen days to return to Canada, creating a log-jam in aircraft deliveries, it was decided to fly them back using what would become known as the Return Ferry Service (RFS).

The only suitable equipment available was the four-engined Consolidated Liberator bomber, which the RAF had just introduced as a long-range maritime patrol aircraft. Beaverbrook persuaded the Air Ministry to release six to be flown by a mix of BOAC and RAF crews. All armament was removed and the bomb bay doors sealed. A wooden floor was fitted to the bomb bay extending to the tail of the aircraft. A passenger oxygen system with twenty individual masks was installed, together with a heating system that would be considered barely adequate for the task. Passengers wore full flying kit with overboots and leather gloves and helmets. The sole concessions to comfort were sleeping bags, rugs and pillows.

The service began on 4 May with the first eastbound aircraft carrying four passengers and 200lb (91kg) of diplomatic mail. Donald Bennett was in command and he took nineteen hours five minutes to fly from St Hubert, Montreal, to Prestwick with an intermediate stop at Gander. The same day, Captain Youell piloted the first westbound Liberator carrying seven passengers, all delivery crews. Bad weather delayed the aircraft at Gander, but its total flying time to Montreal was sixteen hours forty-four minutes.

A description of what it was like to fly as a passenger in an RFS Liberator was provided by Captain George Lothian, a Canadian who after the war would spend two decades as a Trans-Atlantic airline pilot. He recalled his first experience as one of twenty-two passengers, or 'bodies', who boarded the Liberator through a hatch in the lower fuselage near the bomb bay. He took his place on one of the long wooden benches that ran along each side of the fuselage. As there were no windows the passengers 'tended to sit and stare morosely at the far wall'. Then, Lothian recalled: 'Once launched into the sluggish climb, the bodies began to sort themselves out on the floor for the endless 12 to 13 hours to the next stop at Gander. After what seemed 100 years the aircraft finally arrived at Dorval.'

Even so, it had taken the pressure of war to realise a long-held peacetime dream: the year-round trans-Atlantic carriage of passengers and freight by air. The operation quickly settled down to three flights a week. But it soon became clear that more would be needed.

The tempo of the ferry operation increased rapidly in 1941. In the first eleven months the ferry organisation had despatched 266 aircraft from Montreal with all but three arriving safely in the UK, representing a loss rate of just 1.1 per cent. There had been six fatalities, including one in a training accident, but only two crew members had been lost during delivery operations, a loss rate of 0.75 per cent.

Dorval was ready in the autumn 1941 and quickly became the hub of the operation. That summer a further new airfield had become available in a remote location in Labrador. Goose Bay would enable shorter-range aircraft to transit to the UK via Reykjavik, Iceland, and avoid the time-consuming addition of long-range tanks. Soon sixty aircraft a month were passing through this remote terminal. By the end of the war it would be used by more than 24,000 RAF, RCAF and USAAF aircraft as well as the Return Ferry Service. Further intermediate airfields would be established at Greenland under the name Bluie West.

In the second half of 1941 more than 1,350 US-built aircraft were delivered by west coast factories. New routes were established so that aircraft intended for service in North Africa could be delivered via Brazil and West Africa. The US authorities were predicting that 1,000 aircraft per month could be ferried via Canada. At a time when the RAF was seldom reaching its target of 300 land planes a month this prediction was greeted with some scepticism by the British and Canadians. By 1943, however, the South Atlantic route had become as busy with aircraft deliveries as the North Atlantic. The British took responsibility for a return ferry service from West Africa to the US, by which time Ascension Island, a dot of land in the South Atlantic 1,400 miles from Accra, had become an RAF base after an airfield had been blasted out of the lava.

A joint operations room was established at Dorval and manned by US and British personnel. Weather observation stations were set up and a new airway to Europe, code-named Crimson, was planned to go from the US west coast, over Canada via Baffin Island and Greenland to Europe. The US began construction of a new airfield on the west coast of Newfoundland. In 1943 the Canadians got in on the trans-Atlantic act with the launch of a government-supported air service operated by Trans-Canada Airlines. Its equipment comprised a pair of locally built Lancasters with additional fuel tanks in the fuselage. The inaugural flight carried four passengers and 10,000lb of mail.

Also in 1943, RAF Ferry Command was subsumed within a much larger organisation known as Transport Command. Ferry Command became a component of it designated No. 45 (Atlantic Ferry) Group, initially with two wings, one based at Dorval to oversee North Atlantic operations and the other headquartered at Nassau to look after the South Atlantic. A third wing was added later to run Pacific ferry operations.

By the end of the war ATFERO, Ferry Command and Transport Command had delivered more than 9,400 aircraft from North America to the UK. In addition to Hudsons, Catalinas and Liberators, there were also Lockheed Venturas and Lodestars, Douglas DB-7 Bostons, Martin B-26 Marauders and A-30 Baltimores, North American B-25 Mitchells, Douglas C-47 Dakotas and Boeing B-17 Fortresses. More than 1,000 Canadian-built aircraft, including de Havilland Mosquitoes and Avro Lancasters, were also delivered.

With the formation of Transport Command, day-to-day responsibility for the Return Ferry Service passed to BOAC to run until the end of the war. By VJ Day the operation was running one trans-Atlantic service per day and it went on to complete a grand total of 2,000 flights by early 1946. But crossing the Atlantic was still hazardous: in 1941 alone three Liberators were lost with a total of forty-seven returning aircrew members.

Journey times varied but by 1944 RFS Liberators were flying from the UK to Montreal non-stop in thirteen hours thirty minutes and in about twelve hours fifty minutes in the other direction. The organisation had laid the foundations of what Powell later called ‘a fully fledged passenger operation, the first on the North Atlantic’.

The Atlantic Bridge was well and truly open.

Consolidated Liberator

Heavy bomber, maritime patrol aircraft and airliner were just some of the jobs assigned during the Second World War to the highly versatile Liberator.

One of the main reasons for the aircraft's success was its long range. This was due largely to the efficiency of its high aspect ratio wing. By reducing profile drag by up to 25 per cent it brought good fuel consumption and long endurance.

In 1937 freelance aeronautical engineer David R. Davis approached the Consolidated Aircraft Corporation of San Diego with plans for a new type of aerofoil design. At first the company was sceptical, but after tests it selected the Davis wing for a new bomber it was designing for the US Army.

Consolidated had actually been asked to build B-17s under licence, but the company demurred on the grounds that Boeing's design was an old one and that it could do better. Its Model 32 featured a deep fuselage that contained two bomb bays, each one bigger than the B-17's, with doors like a roll-top desk. It also had twin fins and rudders and a distinctive nose 'glasshouse', and it was the first heavy bomber to have a tricycle landing gear.

The Model 32, designated B-24 by the Army, flew for the first time in December 1939. But, although was intended to have superior speed, higher ceiling and longer range than the B-17, initial performance was disappointing. It did, however, improve with the replacement of the mechanically supercharged Pratt and Whitney R-1830 engines by turbo-supercharged units. The relocation of the oil coolers to one side of the engine with the ducting to the turbocharger on the other gave the aircraft its characteristic oval nacelles. More importantly, the change, introduced from the B-24C onwards, increased maximum speed to 310mph (496km).

Although the B-24 is best known for its role in the USAAF's heavy bomber offensive, it first went into service with the RAF. The first six examples, known as LB-30As, were diverted to the Return Ferry Service. Owning to their long range and nineteen-plus hours endurance, they were the only aircraft capable of flying the 2,994-mile (4,790km) journey between Prestwick and Montreal non-stop. 'The Liberator was by no means made for the job and not specially well converted but it had speed and range and gave a minimum of engineering problems,' wrote Air Commodore Taffy Powell, senior air officer to the air officer commanding RAF Transport Command.

The next batch of twenty was delivered in mid-1941 as Liberators Is and, although not considered suitable for the war in Europe because they lacked adequate defensive armament and self-sealing fuel tanks, they were the first to see operational service. The majority were converted for anti-submarine patrol duties, but three served with BOAC displaying civilian registrations. The longer Liberator IIs, equipped with power-operated gun turrets, served as bombers but sixteen were assigned to BOAC. In 1946, these aircraft were returned to the RAF but seven were converted for commercial use. Most were withdrawn the following year with the arrival of more modern airliners.

One Liberator II was modified to serve as Winston Churchill's personal transport. Christened *Commando*, the aircraft featured plush seats and berths, and was later modified with a single fin and rudder. It made many long journeys during the war, but was lost on a training flight in 1945 in the Atlantic near the Azores. Churchill, of course, was not on board.

The US Army also discovered the B-24's value as a transport. The newly established Ferrying Command flew its first trans-Atlantic service on 1 July 1941 when a B-24A piloted by Lieutenant Colonel Caleb V. Haynes took off from Bolling Field, Washington DC, for Prestwick via Montreal and Gander.

The aircraft was also involved in a special diplomatic mission to Moscow in September 1941 when Averill Harriman was despatched as Roosevelt's special envoy to negotiate the terms of the lend-lease agreement with the Soviets. The final 3,150-mile (5,040km) non-stop leg to Moscow followed a route well north of Scandinavia. One of the Liberators returned to the USA via Africa, the South Atlantic and Brazil.

As a bomber the B-24 was less popular with USAAF crews, who considered it more vulnerable than the B-17 to enemy fire, particularly to frontal attack. It was in an attempt to remedy this that a power-operated nose turret mounting two 0.50in (12.7mm machine guns became standard equipment from the B-24G onwards.

Because of its versatility and the demand for it in all theatres of war, the B-24 was built in larger numbers than any other American aircraft. The Ford Motor Company created the world's largest assembly line in a huge new factory at Willow Run, Michigan. Eventually, this facility was turning out 200 complete aircraft per month, peaking at one per hour. More than 18,000 B-24s were built, plus a further 1,000 PB4Ys for the US Navy.

Consolidated B-24J Specification

Crew	11 (pilot, co-pilot, navigator, bombardier, radio operator and six gunners in nose, top, ball and tail turrets and in two waist positions)
Length	67ft 8in (20.6m)
Wingspan	110ft (33.5m)
Height	18ft (5.5m)
Maximum take-off weight	65,000lb (29,500kg)
Power plant	4 Pratt & Whitney R-1830-35 or -41 turbo-supercharged fourteen-cylinder air-cooled radials each generating 1,200hp (900kW) for take-off
Maximum speed	290mph (488kph)
Range	2,100 to 3,700 miles (3,300 to 5,900km)
Service ceiling	28,000ft (8,500m)
Armament	10.50-cal (12.7mm) Browning M2 defensive machine guns plus bomb load of 8,000 to 2,700lb (3,600 to 1,200kg) depending on mission length.

Remembering the Ferry Crews

A memorial to the crews who flew US-built warplanes from Canada to England during the Second World War was erected at Montreal's Dorval Airport. It was later moved to the Dorval Historical Society's building. It reads:

1940–1946
In commemoration to the men of
Royal Air Force Ferry Command
Who under difficult and adverse conditions
Ferried aircraft to the United Kingdom
thus greatly helping to achieve final victory.
This plaque is also dedicated to
the many who gave their lives
in this service for their countries.
We trust Heavenly father their
last flight was successful.

Chapter Thirteen

Just an Entry in a Timetable

Not only did the trans-Atlantic ferry operations help win the war, but they also provided a vital peace dividend.

The conflict had given fresh impetus to the development of navigational aids and terminal infrastructure. Trans-Atlantic air travel had been transformed from an adventure into what was almost a humdrum operation. 'When hostilities began,' observed air transport historian Ron Davies, 'the Empire boats and the Boeing 314s were still headline news; by the end of the war an Atlantic flight was dull routine with every departure no more than an entry in a timetable.'

The expansion of civil aviation could not have happened without the infrastructure left over from the conflict. Airfields such as Gander in Newfoundland and Goose Bay in Labrador, to say nothing of the many landing strips, weather and radio stations built for wartime use in northern Canada, were now available to the airlines. Less tangible but nonetheless important were the methods developed to gather information about the weather and to communicate it to the crews in flight.

Procedures were also put in place before the end of the war to control flights and keep aircraft safely separated in the increasingly busy skies. Canadian historian Dr Carl Christie summed it up when he noted: 'Officers of Ferry Command led in all these areas and bequeathed a lasting legacy to the post-war world.' Even before the conflict ended moves were being made to codify these procedures and enshrine them in an international agreement. In November 1944 representatives of fifty-two states signed the Convention on International Civil Aviation. Better known as the Chicago Convention, it still provides the basis for the operation of commercial air services between nations.

The convention also paved the way for a specialised International Civil Aviation Organisation (ICAO), headquartered in Montreal, to organise and support the intensive international co-operation that the fledgling global air transport network would require. The Chicago conference also led to a set of rules and regulations on air navigation, leading to a common system of air traffic control. The details were worked out region by region, starting with the North Atlantic at a conference held in Dublin in 1946.

That same year, British and American negotiators reached a bilateral agreement regulating the exchange of commercial air services between the two countries. At the first round of talks in Chicago the two countries disagreed strongly about the handling of trans-Atlantic passenger and cargo traffic. The British recognised their weaker position and the two sides agreed to meet again in Bermuda in a bid to hammer out a basis for regular commercial operations, which they eventually managed to do. The resulting Bermuda Agreement established the pattern for another 3,000 similar pacts, but it was denounced by the British in 1977 and a renegotiated treaty replaced it the following year (*see* Chapter 17).

The US had, indeed started, with a big advantage. If any single organisation can be credited with laying the foundations for the American dominance of the trans-Atlantic market in the immediate post-war period that organisation is surely the US Army Air Force's Air Transport Command.

It had grown from Ferrying Command established in 1941 to fly aircraft from factory to trans-Atlantic despatch point into something far beyond that. In four months, Brigadier General Robert Olds built up what one American journalist described as 'an air transit organisation which may dwarf the flying activity of all US airlines put together before this emergency is over'. It was to be a prophetic observation.

After Pearl Harbor it progressed from a domestic ferrying operation into one with responsibility for moving aircraft overseas. By the end of the war it had ferried more than 267,000 aircraft to all parts of the world. In 1942 its responsibilities became even wider when it was renamed Air Transport Command and tasked with the worldwide transportation of personnel and cargo.

Inevitably it drew heavily on the airlines for its most experienced crews and many senior executives accepted commissions to become colonels and majors with Air Transport Command. Experienced pilots, radio operators and other personnel were employed pending the availability of sufficient trained military personnel. The command started with 30,500 personnel and 346 aircraft; by August 1945 this establishment had grown to more than 209,000 personnel and 3,200 aircraft. In July 1945 alone ATC carried 275,000 passengers and 100,000 tons of mail and cargo. It had become an international airline in all but name.

So the US had the organisation and the manpower. It also had the aircraft. There is little doubt that by the end of the war the US enjoyed clear technological leadership in the design and manufacture of long-range airliners. The Douglas DC-4 and the Lockheed Constellation, which were to dominate long-haul air transport in the post-war era, were well in advance of anything else. They had seen service during the war and were now reaching maturity. On the other hand, promising British designs such as the Fairey F.C. 1 and Shorts S. 32 had been abandoned on the outbreak of war, as had a long-range development of the sleek and futuristic de Havilland Albatross.

Perhaps it was these factors that led to the widespread but erroneous belief that the US and UK had agreed that America should build all the transport aircraft needed by the wartime Allies, leaving Britain to concentrate on bombers. The fact is, though, that the end of the conflict did find Britain lacking modern transport aircraft suitable for modern trans-Atlantic operations. Compared with the DC-4 and Constellation, and even Germany's Focke-Wulf Condor, Britain's only airliner, the Avro York developed from the Lancaster, was distinctly crude and unsophisticated.

As for the Condor, Deutsche Luft Hansa had wanted to use it for commercial trans-Atlantic services pre-war and had operated a series of proving flights. In 1938 a Condor flew non-stop from Berlin to New York – and returned non-stop. In a press release, DLH declared: 'As the world's first four-engined, land-based passenger aircraft, with its record time, the Fw 200 indicated the possibilities of transatlantic air travel in the future.' It added: 'The revolutionary aircraft offered room for 26 passengers who travelled in comfortable upholstered seats. For the first time, specially-trained stewardesses were employed onboard and looked after the passengers.' In his dystopian novel *SS-GB* Len Deighton visualised regular London–New York services by Condor.

This achievement had also marked a departure from British and American reliance on flying boats for long over-water routes. Both Pan American and BOAC had operated the big Boeing 314 flying boat during the war. Indeed, when three were ordered for BOAC at £260,000 apiece, this purchase represented the first time a publicly owned British airline had bought foreign aircraft. It was not to be the last.

In 1942–43 the Brabazon Committee studied Britain's post-war airliner needs, but the most successful designs to emerge from its deliberations did not go into service until the 1950s. Interim efforts such as the Avro Tudor and Handley Page Hermes were late in arriving and meant BOAC needed to buy foreign aircraft if it was to remain competitive on the North Atlantic route. It was, after all, still operating the wartime Liberator.

The only choice lay between proven American types. The oldest and least successful was Boeing's Model 307, which embodied experience gained with the B-17. In fact, it combined the bomber's wing and tail with a new fuselage of greatly enlarged and completely circular cross-section with a pressurised thirty-three-seat passenger cabin. Pressure differential was maintained at 2.5lb per sq in and it was powered by four 900hp Wright GR-1820 Cyclones.

The Stratoliner's specification had been formulated in December 1935, but construction did not begin until orders were received in 1937. Four came from Pan American Airways and five from Trans-World Airlines. The first example, intended for Pan American, made its maiden flight on the last day of 1938 but was subsequently lost while being flown by a pilot from another airline. The outbreak of war put a stop to Pan American's plans to operate Stratoliners on a trans-Atlantic mail service.

TWA's examples were bought by the Army for use as C-75 transports to carry VIPs and critical cargo across the Atlantic. By the end of the war they were clearly outmoded but, until the arrival of the Douglas DC-4, they were still the only US-built commercial aircraft able to cross the ocean with a reasonable payload.

Douglas was already the market leader and in terms of sales its DC-4 was to be the most successful of the post-war long-range transports. It originated, however, from a considerably less successful design stemming from a 1935 requirement by United Air Lines for a DC-3 development. American Airlines, Eastern Airlines, Pan American and TWA each contributed $100,000 towards the development costs, but as its complexity rose, Pan American and TWA withdrew their funds in favour of the cheaper Boeing 307.

With a planned day capacity of forty-two passengers, the DC-4, as it was then known, would have double the capacity of the DC-3. It was the first large aircraft with a nosewheel undercarriage and featured innovations such as auxiliary power units, power-boosted flight controls, alternating current electrical system and air conditioning. Cabin pressurisation was also planned for production aircraft.

The wing was similar to the DC-3's with a swept back leading edge, but a departure from the immediate past was the tail, which featured triple fins and rudders to reduce the aircraft's height and enable it to fit existing hangars. After more than two years of construction and testing, the prototype was ready for flight testing. The maiden flight came in May 1938, but further hold-ups and minor modifications delayed type certification for a further year.

The potential customers were not impressed. The complex systems were found to be expensive to maintain and performance was below expectations, especially with an increase in capacity to fifty-two passengers. It was therefore abandoned in favour of a marginally smaller, less-complex design, which was also known as the DC-4. Only one example of the earlier aircraft, retrospectively designated DC-4E (E for 'experimental') was built and that was sold to Japan.

Learning from its mistakes with the DC-4E, Douglas quickly developed the smaller, simpler and entirely conventional DC-4. It had an unpressurised fuselage, a completely new wing and a single fin and rudder. Power came from four Pratt & Whitney R-2000 Twin Wasp engines and seating had been reduced to twenty-eight day passengers. The first flew in February 1942.

By mid-1941 Douglas had orders for sixty aircraft from airlines. But after Pearl Harbor the War Department took over the completed aircraft as well as those still under construction. By January 1946 1,163 C-54s and R5Ds (the army and naval designations for the DC-4) had been built for the United States Army and Navy, many of which passed into civilian ownership.

The aircraft established an outstanding record for reliability and safety. In fact, it was reported that Skymasters completed more than 79,500 oceanic crossings for the loss of only three aircraft, an amazing statistic for the time. The Navy's R5Ds, for example, sustained only one fatal accident in 450,000 flying hours. One C-54 was fitted out for President Roosevelt. Commanded by Lieutenant Colonel H.T. Myers, it completed a number of long oceanic operations including a non-stop flight from London to Washington (3,800 miles or 6,116km) in seventeen hours fifty minutes. It also flew from Washington to Naples (4,200 miles or 6,760km) with two stops in twenty-four hours.

During the war TWA and American Airlines supplied flight crews for Air Transport Command C-54s operating on the North Atlantic. Except in the winter when they flew via the Azores, the trans-Atlantic flights went by way of Newfoundland, Goose Bay or Iceland. Prestwick continued to be the European terminal for these operations. Even after the war military Skymasters continued to be involved in pioneering and development flights. In September 1947 one flew non-stop from Newfoundland to Brize Norton, Oxfordshire, on automatic control from take-off to landing. In 1948 another made a 6,200-mile (9,980km) flight over the North Pole from Fairbanks, Alaska, to Oslo in twenty-two hours.

By then, though, the type had an even more significant flight to its credit. On 23 October 1945 a DC-4 – actually a leased C-54 – had launched the first scheduled commercial trans-Atlantic service. American Overseas Airlines started operations between New York and Bournemouth's Hurn airport, which was then the long-haul terminal for London pending the completion of Heathrow. There were refuelling stops at Gander, Newfoundland and Shannon, Ireland.

This had become possible because in June 1945 the US Civil Aeronautics Board had granted permission to American Overseas and TWA to join Pan American in operating trans-Atlantic services, thereby ending Pan American's monopoly. American Overseas had been established before the war as American Export Airlines and was later acquired by Pan American.

Half-a-dozen long-haul airlines, including Pan American, were to operate DC-4s on their post-war trans-Atlantic services. Yet the use of the DC-4 on these routes was relatively short-lived as the Lockheed Constellation became available in greater numbers. Eventually, American Export began using Constellations and Super Constellations. In August 1947, Pan American began regular scheduled non-stop flights between New York and London using Constellations. The more capable and pressurised DC-6 was also on its way, entering service in 1947.

And by then an improved DC-4 variant was also available, but not from Douglas. The Canadian-built DC-4M combined the basic DC-4 fuselage, wings and tail with four Rolls-Royce Merlin liquid-cooled engines to enable it to fly 40 per cent faster. Later known as the Canadair C-4 Argonaut, this improved DC-4 also offered the benefits of pressurisation.

The first examples, designated DC-4M-2, were, however, unpressurised. Trans-Canada Airlines used them on the Montreal–London route from April 1947 with accommodation for forty first-class or sixty-two economy-class passengers. TCA was not lacking in long-haul experience; in 1943 it had started operating trans-Atlantic air services for VIPs on behalf of the Canadian government. Initially, it used a pair of locally built Lancasters with additional fuel tanks in the fuselage and faired-over nose and tail gun turrets. The inaugural service carried four passengers and 10,000lb (4,536kg) of mail.

A major drawback with the Merlin-powered aircraft, however, was the increased cabin noise compared with the radial-engined DC-4s. But in 1953, by which time the type had been relegated to domestic operations, a TCA engineer evolved a cross-over system for the engine exhausts. This enabled them to discharge outboard and away from the fuselage, reducing the noise level, but not eliminating it.

Although TCA's DC-4-M2s were replaced by Constellations on international and intercontinental routes, Canadian Pacific Airlines operated a variant designated C-4, as did British Overseas Airways Corporation (BOAC). During the 1950s the state-owned British carrier operated twenty-two C-4s on its Empire routes under the name Argonaut. They incorporated as much British-made equipment as possible to reduce the non-Sterling acquisition costs. Deliveries started in March 1949 following the failure of the Avro Tudor, also Merlin-powered. They were also used on BOAC's South American routes before they were withdrawn in 1959–60.

The British government had hoped BOAC would be able to use British-made aircraft on its trans-Atlantic services. But the Tudor, which started as a civilian version of the Lancaster – using the bomber's wing mated to a new circular and pressurised fuselage – was to suffer protracted development problems. These were largely due to escalating demands from the customer and the conflicting priorities of the two government departments involved in the Tudor's acquisition and operation. The Tudor saga, one of the saddest in British commercial aviation history, spoke volumes about the convoluted and bureaucratic nature of official procurement policy.

The demands of war meant that development work proceeded slowly. By the time of the Tudor's first flight in June 1945 the concept had moved some way beyond the original need for an interim type able to compete with the DC-4. Passenger capacity had been doubled but serious handling difficulties revealed themselves during the flight test programme. By the time they were resolved BOAC had lost patience.

British South American Airways, then one of two nationalised British long-haul airlines, was more enthusiastic about the Tudor. It put the improved Mk 4, with its accommodation for up to thirty-two passengers, into service on its Latin American routes in October 1947. But just a few months later, in January 1948, one of these aircraft was lost en route to Bermuda with all its passengers and crew.

The Tudors were grounded pending the outcome of a public inquiry, but as no trace of the aircraft was ever found there could be no definitive conclusion about the cause of its loss. Donald Bennett, who was now BSAA's chief executive, insisted that sabotage had been the cause. Indeed, he was so outspoken in his belief that he was sacked. As there seemed no particular reason to keep the aircraft grounded, the Tudors went back into service. But in January 1949 another disappeared en route from Bermuda to Jamaica. Again, no trace of the aircraft was ever found, so its loss could not be attributed to any particular cause. That, however, proved to be the end of this ill-starred aircraft's front-line career: it never operated on the North Atlantic route as it had been designed to do.

It was while this dismal affair was being played out that BOAC sought government permission to spend some of Britain's dollar reserves on a modest fleet of Constellations. Their arrival meant the airline could retire the Liberators it had been using during the war. Future best-selling novelist David Beaty was a Liberator first officer and he recalled how the aircraft carried mostly mail and freight with the occasional staff passenger, who was obliged to bed down 'on top of the mail bags in the icy, draughty fuselage'.

On his first trip from Montreal to London via Gander, Beaty was introduced to the 'buttercups', the yellow cardboard coffee cups that doubled as receptacles for cigarette ends after surreptitious smokes. When full these cups were tossed out of the window. As a result, it was said, there was no need for navigation because you simply followed the trail of discarded cups on what became known as the Buttercup Route. Over the ocean, though, things could get pretty rough for Liberator passengers, especially during an Atlantic storm. Beaty recalled: 'Roller coasting their way blindly through heavy turbulence, passengers were as sick as dogs.'

To reduce the dollar expenditure involved, BOAC was allowed only five Constellations that had previously been operated by the USAAF. In some quarters the nationalised carrier was accused of letting the side down, but the purchase meant that BOAC was able to launch the first regular British services to New York in July 1946. By that time the airline had completed more than 2,000 Atlantic crossings. More than 20,000 passengers had been carried together with nearly 4 million lb (1.8 million kg) of freight and mail. Forty-five captains had each made more than 100 crossings and by July 1946 twelve were flying the Constellation.

Captain O.P. Jones was in command of BOAC's first scheduled east–west Constellation service on 1 July 1946. A fortnight earlier *Flight*'s Maurice Smith had accompanied Captain W.L. Stewart, a former Return Ferry Service pilot, on his 161st crossing during a proving flight from Heathrow to New York's La Guardia airport via Shannon and Gander aboard *Bangor II* (G-AHEL). Departure was postponed twice, once due to bad weather and then because of cabin pressurisation trouble. The aircraft took off the following day without pressurisation, limiting it to flying below 10,000ft (3,000m). There was further discomfort at passport control, which in the days before permanent terminal buildings was conducted in a draughty tent.

On board, Smith found the aircraft 'quite comfortable and roomy' with adequate seat adjustment. Vision through the cabin windows was found to be good and the cabin's uniform neutral colour scheme 'restful to the eyes'. The same could not, though, be said of the roof lighting, which produced 'a number of bright and disturbing spots of light'.

The cockpit layout, Smith reported, was 'characteristically systematic', with most of the engine controls grouped before the flight engineer and the essential dials, switches and levers duplicated for the pilots. Ahead of each control column was the blind flying panel and, just outboard, the blind landing indicators. In the middle were the duplicate engine boost and revolution counters and the Sperry auto-pilot controls. In a central overhead panel were the radio compasses, VHF and HF radio controls and, beside the captain's seat on the port side, the electrical system switches. The flight engineer's station was behind the two pilots, as was the radio operator's position with the navigator's table and crew rest room further aft.

The radio compass was used for trans-Atlantic navigation and also for wireless telegraphy fixes. Loran and astral navigation was also used, although Loran suffered a 'dead' area in mid-ocean. Meteorological information was based on the hourly reports sent by all trans-Atlantic aircraft. Using dead-reckoning alone the maximum track error seldom exceeded 30 miles (48km).

The return flight, a week later, was commanded by Captain L.V. Messenger. In his report, Smith noted that it:

> seemed, to bring home more forcibly what travel by air can mean: lunch on Sunday in New York, take-off, read the newspaper, have tea and Gander is seen below. Take off again, put watch on five-and-a-half hours and it is bed time. Sleep accounts for the over-ocean period and breakfast in Ireland is the next event. Final take-off follows and then two hours airborne plus one in a car and one can be in London just 24 hours after leaving New York.

BOAC's first L-049 Constellations were followed by further acquisitions. In 1948 the airline was able to buy five of the improved Model 749 from Aer Lingus. They had become available because of a change of plan by the Irish carrier. The resulting deal was particularly attractive to the British government because it meant nearly-new aircraft could be acquired without the expenditure of precious dollars.

There was a brief hiatus when Constellations were grounded for six weeks following an accident to other operators' aircraft, which cost BOAC more than £120,000 in lost revenue. But Constellations remained in its service on the North Atlantic until early 1950, when the type was switched to other routes.

By the end of the 1940s eleven airlines – Pan American, BOAC, American Export, TCA, TWA, KLM Royal Dutch, Air France, Scandinavian, SABENA, Iceland's Loftleider and Swissair – were offering trans-Atlantic services. They had settled down to provide reliable and comfortable operations with their Constellations or DC-4s and collectively were carrying around 100,000 passengers a year.

None had an outstanding equipment advantage, but then, according to airline historian Ron Davies, the 'game of one-upmanship began'. The stakes were first raised in 1949 with the introduction of the Boeing Stratocruiser, operated by Pan American, BOAC and American Overseas. Meanwhile, Douglas had developed the pressurised DC-6 from the DC-4.

Douglas had produced it to compete with the Constellation. The launch customer was United Airlines, but in April 1947 the type was grounded for four months following a number of in-flight fires. It was, however, considered marginally more economical to operate than the Constellation and easier to maintain. A total of 704 was produced, including 288 of the stretched DC-6Bs, of which Pan American operated forty-five examples. The DC-6 was introduced to the North Atlantic in July 1950 when the Italian airline IAI (later merged with Alitalia) launched its Rome–New York service.

The Stratocruiser was developed from the XC-97 military freighter, which incorporated features from the B-29 bomber, notably its wings, tail and undercarriage. Although the lower part of the bomber's fuselage was retained – with cargo and baggage holds replacing the bomb bays – the new design added an upper lobe of much larger diameter, which was to produce the distinctive inverted figure of eight cross-section.

Boeing wanted to offer a civil variant, but the airlines reacted cautiously. They appreciated the promised performance, but were concerned about the complexity, size and cost. Boeing, however, decided to build fifty examples at its own expense and Pan American set the ball rolling by ordering twenty. The order book peaked at fifty-five, six of which were delivered to BOAC in 1949.

That year Pan American introduced its Stratocruisers on the New York–Bermuda route, followed in June by New York–London services. On its President service, Pan American's Stratocruisers generally carried sixty-one passengers eastbound and fifty-six westbound, with a stop at Gander. Once a week there was a surcharged service on aircraft configured with twenty-eight berths and seventeen 'slumberette' seats.

BOAC followed Pan American in December 1949 and its Monarch trans-Atlantic services were operated daily from the following February. The blunt-nosed Boeing set new standards for luxury travel. Particularly popular was the downstairs bar accessed by a spiral staircase. So too were the spacious bunks. Both were worthwhile features on flights that could take twenty hours, longer if there was an engine failure or other unscheduled delay.

Despite the modern equipment, icing and mechanical troubles were still to be expected. Sometimes you got both. Captain Ian Jeffery recalled a runaway propeller during a westbound Constellation trip. 'We managed to get the propeller stopped and feathered but we were icing up at the time and discovered that we had an 80-knot tailwind,' he told the author. 'We had no option but to carry on to Gander, five and a half hours on three engines, loaded up with ice. It took nine hours fifty-five minutes to get from Prestwick to Gander.'

Mechanical trouble was something crews lived with in those days of hugely complex piston engines. Former BOAC Stratocruiser captain Tony Spooner recalled how he and his colleagues had accumulated 'scores, if not hundreds of three-engined hours on Strats'. And it could get worse: 'Some of us even collected a few two-engined hours.' These,' he added, 'were *not* so comfortable.'

But the Strat lacked the endurance to fly non-stop between London and New York or Montreal. Often the journey required stops at Prestwick, Shannon and Gander or Goose Bay. Sometimes they would go via Keflavik, Sydney (Nova Scotia) or Moncton (New Brunswick). The route was determined by the weather and it was up to captains to decide which one to take. 'Captains soon became meteorologists first and pilots second,' Spooner recalled.

Ian Jeffery preferred the Stratocruiser to the Constellation because of its more spacious flight deck, but acknowledged that it struggled to rise above the weather. He told the author: 'You might, when you'd burned off enough fuel, but after take-off your chances of getting above the weather were pretty well zilch. With a full load of fuel you were lucky to get above 13,000 ft [3,900 m].'

The competition between Douglas and Lockheed intensified with the introduction of the DC-7 and the Super Constellation. The DC-7 represented the ultimate development of a line of piston engine airliners that had started in the early 1930s with the DC-2. It was essentially a stretched DC-4/DC-6 and, while the baseline version was used only for domestic operations, the DC-7B and -7C were intended for long-haul services. In fact, the -7C was designed for regular non-stop trans-Atlantic operations with sixty-four to ninety-five passengers. All three variants were powered by Wright turbo-compound R-3350 engines, which developed 3,250hp to 3,400hp. A total of 121 -7Cs were delivered to thirteen airlines, one of which was BOAC, which bought ten to fill the gap caused by delays to the Bristol Britannia turboprop programme.

The Super Constellation appeared in 1953 and was introduced to the North Atlantic in its Model 1049C form by KLM that August. The Dutch flag carrier was the first to schedule non-stop flights between New York to Amsterdam. Air France followed in November and TCA in May 1954.

BOAC, however, was having a difficult year with an acute shortage of capacity following the grounding of the de Havilland Comets. Despite the acquisition of additional Constellations and Stratocruisers, the British carrier's share of the market fell and was not restored until the second half of the decade.

In June 1955 the revived Lufthansa opened its first service to New York with the latest iteration of the Super Constellation, the 1049G. TWA (the initials now stood for Trans World Airlines) put its 'Super Gs' into service on the North Atlantic in November and by the following spring had built the operation up to fifty crossings a week.

Pan American, meanwhile, was supplementing its Stratocruiser President operation with the Rainbow service using DC-6Bs from May 1952 and DC-7Bs from June 1955. A year later, Pan American launched the DC-7C Seven Seas, scheduling around seventy-five flights a week. In 1956 it became the first airline to carry 200,000 passengers over the North Atlantic in a single year. In fact, the carrier remained top dog throughout the 1950s. TWA's response was to launch the Jetstream service using the Lockheed Model 1649A Starliner.

With the growth in importance of the US west coast market and the availability of long-range airliners, direct flights between cities such as Los Angeles and San Francisco and European destinations now became possible. The quickest route was a great circle over the North Pole.

Non-stop trans-polar flights from Europe to the US Pacific Coast had been pioneered in June 1937 by Soviet pilot Valery Chkalov. His flight, from Moscow to Vancouver via the North Pole in a Tupolev ANT-25 single engine aircraft, took sixty-three hours to complete. The distance covered was 5,475 miles (8,811km).

But it was to be another seventeen years before the route was used by scheduled airliners. In November 1954 Scandinavian Airlines System inaugurated its Copenhagen to Los Angeles service via Greenland and Winnipeg using DC-6B aircraft. The journey time was twenty hours. Canadian Pacific Airlines was next with a Vancouver–Amsterdam service, stopping at Edmonton and Greenland, and Pan American followed with a Los Angeles–London operation via San Francisco using DC-7Cs. Not to be outdone, TWA launched a similar service in October 1957 using Starliners.

With their ability to fly non-stop between New York and any European capital city, the DC-7Cs and Starliners represented the ultimate expression of the piston-engined airliner. But before the turbine-powered machines rendered them obsolete they had collectively carried more than a million passengers across the North Atlantic.

Indeed, Atlantic passenger volumes grew spectacularly during the 1950s. Pan American started the decade carrying 69,000 passengers but this number had grown to 336,000 in 1959. TWA carried 66,000 passengers in 1950 and 146,000 by 1959, by which time it had been overtaken by BOAC. The British airline had moved from 36,000 passengers in 1950 to 211,000 by 1959 and was now out-carrying all its trans-Atlantic rivals apart from Pan American.

This was largely due to the introduction of turbine power. The stage was now set for an even greater technological leap forward.

The First Trans-Atlantic Landplane Operations

Airline	Date	Route
American Export (DC-4)	24 Oct 45	New York–Hurn via Gander, Shannon
Pan American (L-049)	3 Feb 46	New York–Bermuda
TWA L-049)	5 Feb 46	New York–Paris via Gander, Shannon
Pan American (L-049)	11 Feb 46	New York–Hurn via Gander, Shannon
KLM (DC-4)	21 May 46	Amsterdam–New York via Prestwick, Gander
Air France (L-049)	24 Jun 46	Paris–New York via Shannon, Gander
BOAC (L-049)	1 Jul 46	London–New York via Shannon, Gander
SAS (DC-4)	16 Sep 46	Stockholm/Copenhagen–New York via Prestwick, Gander

The Lockheed Constellation

For most of the post-war era the Lockheed Constellation remained in a class of its own, even though it was outsold by its Douglas rivals. It was around 80mph (128kph) faster than the DC-4, offered sixty seats against forty-four at the same pitch and was pressurised.

Production started in 1942 and terminated in 1958, by which time the US armed forces had bought 331 and the airlines 524. Over those sixteen years it grew from an 82,000lb (37,000kg) aircraft with 2,200hp engines to a 160,000lb (7,250kg) machine with 3,400hp engines. The original wingspan of 123ft (37.5m) and length of just over 95ft (29m) were raised, respectively, to 150ft (45.7m) and just over 116ft (35.4m).

Howard Hughes was already a Lockheed customer when, in 1939, he approached the company with proposals for a new airliner for TWA, the airline he had just acquired. Lockheed executives showed him the Excalibur concept on which they had been working, but Hughes wanted something better – nothing less than 'the airliner of the future'.

That meant a cruising speed of at least 300mph (480kph), accommodation for up to sixty passengers and the ability to fly non-stop across the USA or even from New York to London. These demands were far ahead of anything then available or even on any drawing board.

But the challenge had been issued and Lockheed's response was the Model 049 Constellation, which first flew in 1943.

It was quickly impressed into Army Air Force service as the C-69. In fact, all examples built up to the end of the war were bought by the Army. Afterwards Lockheed bought back as many as were available and, together with those C-69s still on the production line, converted them back to model 049 specification.

A major drawback, however, was the Wright R-3350 Cyclone engine, whose teething troubles included overheating and even fires. These issues were still unresolved by the end of the war and several incidents in 1945–46 caused Constellations to be grounded for a time, although problems with the complex engines continued to afflict them.

In late 1946 the 049 was succeeded by the Model 649, which featured strengthened internal wing structure, landing gear and brakes as well as better cabin heating, cooling and ventilation. Cruising speed rose to 327mph (526kph), or 14mph (23kph) above the 049. The succeeding Model 749 was designed to meet airline demands for more range, and fuel capacity raised by 1,555gal (5,886lit) boosted this with full payload to 2,600 miles (4,184km).

The first stretched aircraft was the 1049 Super Constellation, which was 18ft 4in (5.6m) longer than previous models. It also had more powerful turbo-compound engines and carried 6,550gal (23,974lit) of fuel with an optional wing tank holding another 730gal (2,763lit).

TWA inaugurated non-stop 1049 services between Los Angeles and New York on 19 October 1953. At first the type suffered problems with exhaust gasses flaming excessively but once these were resolved it went on to become highly popular and several sub-variants were developed. The 1049G featured wing tanks to raise range by 700 miles (1,120km), while the final variant, the 1049H, could be converted from airliner to cargo carrier and vice versa in a matter of hours.

In May 1955 Lockheed started work on the ultimate Connie. The Model 1649 Starliner featured a longer, narrower wing that nearly doubled the Model 049's fuel capacity to 9,278gal (35,121lit) and more than doubled its range with maximum payload to more than 5,400 miles (8,690km). The Starliner offered high levels of passenger comfort. Setting the engines further from the cabin in the longer wings reduced internal noise. There was improved cabin temperature control and ventilation, and seats were fully reclining for extra comfort on long flights.

Lockheed Model 1049G Specification

Flight crew	five
Passengers	62–95 with 109 in high-density configuration
Length	116ft 2in (35.42m)
Wingspan	126ft 2in (38.47m)
Height	24ft 9in (7.54m)
Maximum take-off weight	137,500lb (62,370kg)
Power plant	four Wright R-3350-DA3 turbo-compound eighteen-cylinder supercharged radials each developing 3,250hp (2,424kW)
Cruising speed	340mph (547kph) at 22,600ft (6,900m)
Range	5,400 miles (8,700km)
Service ceiling	24,000ft (7,620m)

Chapter Fourteen

When Britannia Ruled

For ten brief months in the late 1950s there was only one way to fly the North Atlantic. If you wanted the fastest, the most comfortable and the most stylish way of travelling between London and New York you had to go on a British aircraft.

The dream that had started with the Brabazon Committee and its proposals for post-war British airliners had finally been realised. Suddenly, the Super Constellations and DC-7s were obsolete. Pan American, TWA and the other airlines were left wondering where the traffic had gone.

There was rivalry, but for once the American airlines were not involved. Going into the turbine age, the honour of operating the first trans-Atlantic services with the new Bristol Britannia turboprop lay between the British Overseas Airways Corporation and the Israeli airline El Al.

To many observers, the starting gun in the race to be first was fired on 6 December 1957. That day's New York *Herald Tribune* carried an eye-catching full-page advertisement placed by El Al. It showed a photograph of the sea with a simple message: 'Starting 23 December the Atlantic will be 20 per cent smaller.' It was the next line that caused the biggest stir: 'Watch for the inauguration of the first jet prop service across the Atlantic, introducing the Bristol Britannia.'

The next day one of El Al's three Britannia 312s made the fastest ever crossing by a commercial airliner from New York to London. It took eight hours three minutes. Until then BOAC had been talking about launching its turboprop Britannia on the North Atlantic, 'within the first three months of 1958'. The airline now decided to advance the date to 19 December, although some maintained this had always been its real intention.

El Al's response was a proving flight from New York to Tel Aviv that, it was claimed, broke six world records. Among them was completing the 6,100-mile (9,760km) non-stop journey in fourteen hours fifty-seven minutes. The Britannia had flown further and faster than any other commercial airliner in history.

BOAC had introduced the Britannia in its short-range Series 102 guise on the London–Johannesburg route on 1 February 1957. The three times weekly services involved three stops en route and took nearly twenty-three hours. On 16 December BOAC introduced the long-range Series 312 on the North Atlantic.

Promoted as the Whispering Giant, the Britannia became the first turbine-powered airliner to operate scheduled passenger services on the London–New York route when G-AOVC left Heathrow commanded by Captain A. Meagher. It was scheduled to fly to New York's Idlewild airport via Prestwick, the southern tip of Greenland, Gander and Boston. Even though the aircraft encountered strong crosswinds, and was diverted to avoid the worst of them, 'VC's air speed averaged 350mph (560kph) over the 3,750-mile (6,000km) journey. Conversely, tailwinds on the return journey two days later enabled the aircraft to average 405mph (648kph).

BOAC operated its trans-Atlantic Britannias in an all first-class configuration, with the luxurious Majestic class cabin furnished for just fifty-two passengers. There were twenty-six fully reclining sleeper seats and a similar number of conventional recliners. At the rear of the cabin were two spacious dressing rooms with specially toned lighting. There were three separate toilets. According to the size of the load carried, two toilets could be made available for women and one for men – or vice versa – by use of an adjustable partition between the dressing rooms.

Because of the light loads carried, BOAC was able to make most of the crossings non-stop, although there were times when strong headwinds meant refuelling stops at either Goose Bay or Gander. The east-bound scheduled time was set at twelve hours, although it was frequently accomplished in less; on one occasion in 1958 it took just nine hours thirty minutes. The westbound service was scheduled to take eleven hours.

Initially, BOAC operated just one Britannia service a week on the New York route. The airline had hoped that it would have a three-year lead over its competitors, but delays in receiving the aircraft reduced this to just ten months. Yet the first turboprop still represented a significant milestone in trans-Atlantic operations. 'It was,' observed historian Ron Davies, 'the first time that a British-built aircraft had flown in regular, sustained Atlantic service; and the Britannia was the first genuine non-stop Atlantic aircraft in both directions.'

It had been a long time coming. 'Since the first meeting of the Brabazon Committee in 1943,' the journal *Flight* observed in December 1957, 'this country has been planning for the day when a British aircraft flown by a British airline would operate non-stop passenger services between London and New York.'

The Britannia had been the final expression of a concept originally formulated by a committee formed to consider the kind of airliners British airlines would need after the war to keep the Union Flag flying, particularly on the prestigious North Atlantic route (*see* box). The committee's ideas were taken up by the Ministry of Supply, which wanted a medium-range airliner suitable for BOAC's Empire routes.

This aircraft had originally been intended to have piston engines, but by 1949 the decision had been taken to substitute Bristol's Proteus turboprop units then under development. The requirement was also changed to a long-range airliner capable of trans-Atlantic operation. The prototype first flew in August 1952.

As was not uncommon in those pioneering days, development was delayed for a number of reasons, not least problems with the complex Proteus engines. Icing in certain conditions was encountered during route-proving operations and took a year to resolve. Once in service, however, the turboprop proved popular with passengers, who liked its smoothness as well as its speed. This was reflected in the volumes of traffic carried. In May 1958 the New York service was extended to San Francisco, to mark the first British service to the US west coast.

Israel's national airline El Al was also quick to spot the potential of turbine power. It became the second Britannia customer and followed BOAC on the North Atlantic with the launch of Tel Aviv–New York operations. Services were quickly stepped up to five a week and the airline doubled its share of the trans-Atlantic market within a year, reflecting the Britannia's ability to operate a non-stop service in all but the most severe winter weather. El Al was quick to exploit this attribute in its advertising. The slogan, 'no Goose, no Gander', was a reference to the diversionary airports that were then much in use and which passengers could avoid by travelling by Britannia.

Inevitably, the Britannia's introduction was overshadowed by the arrival of the pure-jet airliners. For it was just ten months later, on Saturday, 4 October 1958, that BOAC pulled off a dazzling public relations triumph when two of its sleek, newly delivered de Havilland Comet 4s took off from opposite ends of the London–New York route to inaugurate the world's first trans-Atlantic commercial jet service.

G-APDC had left London's Heathrow airport, while G-APDB made the journey in the opposite direction from Idlewild to make it a stunning double first. The westbound Comet took ten hours twenty-two minutes to complete the flight to New York with a one-hour refuelling stop at Gander, while its eastbound counterpart reached London in a record-setting six hours eleven minutes that halved the normal journey time.

As with the earlier Britannia launch, there had been a contest between two airlines for the honour of being first. This time, BOAC was up against Pan

American with its US-built Boeing 707s. Britain had determined to be first again, but its lead in jet airliner technology had shrunk from several years to just twenty-two days.

Eight years earlier the Comet was undergoing flight testing while America's first jet airliner was little more than a concept under consideration by Boeing's engineers. In September 1950 a party of the company's senior executives was at the Farnborough air show to watch the Comet, which would become the world's first commercial jetliner in less than two years' time, being put through its paces.

To company president Bill Allen it 'appears to be a fine airplane' and he repeated this opinion over dinner that evening. The Boeing party discussed the possibility of building a rival, perhaps with podded rather than buried engines and more pronounced wing sweep-back. Allen, though, kept coming back to the cost involved and remained non-committal about it. Later, when European airline bosses asked him about Boeing's plans for a jet transport, all he would say was: 'We're reviewing it.' His private view was that building a jet the US airlines were not then clamouring for might just be pouring millions of dollars down the drain.

Within two years, though, Allen's view would change. Boeing now realised that the B-47 bombers it was building for the US Air Force would need tankers able to sustain them and, even though they might not be aware of it, the airlines would need jets before too long. In fact, Boeing had been thinking about jet transports even before Allen's visit to Farnborough. One team was working on an improved Model 367 KC-97 tanker, the military equivalent of the Stratocruiser. At one stage this team would propose what was essentially a jet-powered KC-97 with bluff, double-bubble fuselage, mildly swept wing and four engines mounted in two underwing pylons. This was known as the 367-64. Another team was considering an airliner based on the B-47 bomber with shoulder-mounted wing, paired engines in pods and bicycle landing gear.

Britain had been thinking about commercial jets since the early 1940s. Less than two years after the first British jet had flown, the Brabazon Committee was considering a jet-powered mail carrier with North Atlantic capability. At the same time, engineers at the de Havilland company, home of high-speed aircraft such as the Mosquito, were brainstorming advanced airliner concepts. One of the ideas they came up with was an unconventional tail-less design with three rear-mounted jet engines.

Geoffrey de Havilland was the only permanent manufacturer represented on the committee. His company not only built aircraft but, through the work of Major Frank Halford, was also a leading pioneer of jet engine technology, making it well-suited to the task of designing a jet airliner. In any case, by 1945 de Havilland had become convinced that jet propulsion represented the future.

The government saw these developments as a way of using Britain's lead in jet engine technology to leapfrog the American manufacturers whose advanced piston engine designs were set to dominate post-war commercial aviation. The result was the DH 106. With its moderately swept-back wings and engines buried in the wing roots, it was rather more conventional than some of the early concepts, but it still represented a massive technological leap forward.

From its method of construction to its de Havilland Ghost engines – the first practical commercial jet power plants – the Comet was pushing back the frontiers of technology to an extent not seen before. And with great rapidity too: the prototype made its maiden flight, commanded by de Havilland's chief test pilot John Cunningham, on 27 July 1949. A year later a second aircraft had joined the flight test programme and it was becoming clear that the Comet was capable of cruising 200mph (320kph) faster than the Constellations and DC-6s then in service. In October 1949, the prototype whistled from London to Tripoli, Libya, 1,468 miles (2,350km) away, in three hours twenty-three minutes, an average speed of 434mph (695km).

In January 1952 the Comet received its Certificate of Airworthiness and on 2 May BOAC put the world's first jet airliner into commercial service between London and Johannesburg. There were intermediate stops in Rome, Cairo, Khartoum, Entebbe and Livingstone, yet the jet was so fast it could be halfway back to London before a DC-6B had reached Johannesburg from Heathrow.

The Comet was indeed proving to be a fine airplane. But the new standards it was setting were not available to trans-Atlantic passengers because the Comet 1 lacked the range for trans-oceanic journeys.

Boeing was still finding customers somewhat unresponsive to the idea of jets. The air force maintained it lacked the budget to buy jet tankers, while the airlines had just re-equipped with the latest piston-engined airliners and did not want expensive jets. This placed the company on the verge of a critical decision. Bill Allen believed an airborne demonstrator would generate interest in jets in both military and commercial markets. He managed to persuade his board to back him by investing a sum equivalent to two-thirds of the company's net profit for the post-war years.

Company historian Mike Lombardi told the author:

> Bill Allen and his team did feel the time was right but unfortunately they could not find customers who agreed – including the US Air Force. This is what makes his decision to invest all of the company's resources into building the Dash 80 one of the most important in business history. The leadership at Boeing understood that the future was in jets and staked the future of the company on that belief.

The Comet, meanwhile, seemed to be going from strength to strength. In August 1952, three months after the inaugural London–Johannesburg operation, BOAC launched services to Colombo, Ceylon (now Sri Lanka). This was followed two months later by a service to Singapore and then, in April 1953, the first Comet flew into Tokyo. In less than two years, BOAC had created the first ever jet airline network. By early 1954 thirty cities in the Far East were being served by Comets.

There had been accidents during this period, however. An in-flight break-up near Calcutta was attributed to abnormal stresses encountered in a violent thunderstorm. Take-off accidents at Rome and Karachi airports were put down to pilots' unfamiliarity with the new jets. But when, in January 1954, a Comet plunged into the Mediterranean near the island of Elba there appeared to be no reasonable explanation. BOAC grounded all its jets for an exhaustive inspection, but they resumed flying two months later when nothing untoward was found. Many observers wondered if sabotage was responsible.

This, however, was forgotten in April when another Comet was lost in similar circumstances in the Mediterranean near Sicily. This time the Comet's Certificate of Airworthiness was withdrawn and a reappraisal of the entire structural integrity of the world's first jet airliner was begun. The Royal Navy initiated an extensive salvage operation to retrieve as much of the wreckage of the two Comets as possible. Eventually, enough was recovered to enable investigators to prove that a hitherto unknown phenomenon had been responsible for the accidents. The finding that metal fatigue through repeated bending during the pressurisation cycles had been responsible revolutionised aeronautical design.

On 14 May 1954, a month after the second Comet loss over the Mediterranean, the Boeing Model 367-80 prototype was rolled out at its Renton, Washington, plant. Resplendent in a cheerful custard yellow and chocolate brown colour scheme, it was intended to be a proof-of-concept vehicle rather than a production prototype. It would not be type-certificated. But, according to Joe Sutter, then head of the project's aerodynamic team:

> The most important thing about the Boeing 367-80 was that it truly defined the jetliner. In this very broad sense it's the ancestor of all the commercial jet transports we take for granted, not just the Boeing 707 or later Boeing jets. The Dash 80 is the granddaddy of them all, right up to the latest Airbus products or the Boeing 787 Dreamliner.

The Dash 80 flew for the first time on 15 July 1954 with chief test pilot, Alvin M 'Tex' Johnston, in command. To Bill Allen the moment it lifted off was 'the most beautiful first flight take-off I ever witnessed'. After one hour twenty-four minutes the prototype landed at Boeing Field well within the 10,000ft (3,076m) runway. Johnston said it had 'flown like a bird, only faster'.

As the bugs were being ironed out, Boeing executives sweated on finding buyers for their revolutionary and expensive aeroplane. Then, on 1 September 1954, the USAF ordered a first batch of twenty-nine KC-135A tanker/transports based on the Dash 80 for Strategic Air Command. There was much relief in the Boeing board room, but civil sales were still hard to find: there was plenty of interest, but none of the airlines seemed willing to place an order. The jets were still seen as an enormous gamble.

To counter public nervousness, Boeing used the Dash 80 prototype for a series of press and customer flights. An advertising campaign stressed the comfort and safety offered by jets. The company also played on US airline fears that, with a revised version of the Comet based on lessons learned from the accidents, the British might yet retain their lead in commercial jets.

Although initially sceptical, the Douglas company also decided to press ahead with a jet and it launched the DC-8 in July 1955. That October, Pan American stunned the industry by ordering forty-five new jet liners: twenty-five DC-8s and twenty Boeing 707s. Each had double the capacity of current propeller-driven airliners with potential to make non-stop trans-Atlantic crossings. Pan American had previously ordered Comets, but cancelled them after the 1954 accidents. Accordingly, noted Ron Davies, the October 1955 announcement 'read like an obituary in the board rooms of Britain's aircraft industry'.

The loss of the Comets and ensuing worldwide publicity could have killed off the world's first jetliner, yet an improved version was on the way. The Comet 4 was announced in March 1955 and, just as Boeing had feared, it incorporated lessons learned from the accident investigation. It was very different from the original Comet 1. For a start, it was 20 per cent bigger and 12 per cent faster. More efficient axial-flow Rolls-Royce Avon engines replaced the original de Havilland Ghost centrifugal power plants and offered 135 per cent more power. The Comet 4 could carry 125 per cent more passengers over a 115 per cent greater range compared with the Comet 1.

Early in 1958 Pan American announced plans to begin trans-Atlantic 707 services that October. BOAC, which had not intended to use its Comet 4s on the route, declined to be drawn on speculation that it was changing its mind. How much pressure was coming from the government was not clear; the chance of beating Pan Am and the Boeing 707 to become the first trans-Atlantic jet airline and salvage some of Britain's aeronautical prestige was too good to miss.

History shows that the Comet did indeed beat the 707. The margin was a modest one, but a point had been made. And the Comet quickly become popular with passengers. One of the first American ladies to fly in it was captivated by the cabin décor. 'Isn't this the dreamiest?' was the reaction noted by *Flight*'s Mike Ramsden, the de Havilland apprentice turned journalist.

He flew on one of the early trans-Atlantic services and described the Comet's interior as 'a work of art, a masterpiece of design so subtle that print can but inadequately convey the pleasure which it gives'. Ramsden commented in particular on the variations in colour between the linen headrests and cushions and the upholstery. The seats were pronounced 'most comfortable', but high standards were expected: BOAC's Monarch service had become a byword on the north Atlantic. 'Its refinements could not have had a more appropriate setting than the aesthetically pleasing and physically tranquil cabin of the Comet,' Ramsden added.

All this came at a price, of course. Passengers who chose de luxe travel occupied sixteen seats in the forward – and quietest – section of the cabin, which enjoyed a seat pitch of 56in (142cm). The aft cabin accommodated thirty-two first-class passengers who paid £155 for the round trip. The de luxe surcharge was £18, almost double the UK average weekly wage at the time.

Pan American had to content itself with being the second airline to operate trans-Atlantic jets. On 26 October 1958, *Clipper America* inaugurated Boeing 707 services from New York to Paris. Before departure, the commander, Captain Sam Miller, briefed the crew of Flight 114. Also attending were the first officer, Captain Waldo Lynch, the navigator Captain A.O. Powell and flight engineer Jim Etchison. Flight attendant Jay Koren recalled: 'Capt Miller quizzed each of us on emergency equipment and procedures: "Where are the life rafts? How many are there? How many passengers should you direct to each?"'

At 1800, an hour before the scheduled departure time, the crew bus delivered the crew to the foot of the forward loading stairs. It was drizzling lightly and Koren thought the military band, assembled on the tarmac 'to lend the occasion appropriate pomp and circumstance, looked damp and uncomfortable'. By the time the cabin crew boarded, the flight crew were going through their checklists. A trio of *Life* magazine photographers were busy stowing their lights and cameras.

New York's Mayor Wagner and Pan Am president Juan Trippe lent their presence and words to the momentous event. A ribbon was cut and the passengers eagerly ascended the steps. Last to board was actress Greer Garson, who stole the scene. Delivered to the base of the steps in a Rolls-Royce, she turned at the doorway and, displaying her Oscar-winning *Mrs Miniver* smile, tossed an enormous bouquet of pink roses to the crowd below as photographers lit up the night with a frenzy of flash bulbs.

Koren recalled the sensation of power as the aircraft accelerated towards its take-off speed:

> The powerful force pressed our backs into the thinlypadded bulkhead behind us. Even more startling was the unexpected vibration and violent roar of the jet engines as we gathered speed for our leap up into the night. We grasped hands and stared wide-eyed at one another in disbelief. Where is that vibration free, quiet-as-a-whisper ambiance the airline ads have been touting? We discovered why the first-class section is now located in the front. Just opposite to piston-engine aircraft – where the cabin becomes quieter toward the rear – we were seated in the noisiest spot in a jet.

But it all changed once the cruising altitude was reached. 'The vibration disappeared completely and the engine roar subsided to little more than a gentle hum,' Koren wrote later. 'The lights came up, seat belt signs went out, and it was time for champagne and canapés. An exhilarating aura of festive exuberance prevailed.'

There was a refuelling stop at Gander and, after seven hours' flying time, the aircraft landed in Paris, where Greer Garson once again stole the show. For the return trip, Koren recalled, 'total elapsed time, Paris to New York, including one fuel stop in Keflavik, Iceland, was nine hours versus 20 on my Stratocruiser crossing, the same flight number, the previous week'.

Normally there were six cabin crew members on board Pan Am's DC-8s and 707s – two pursers and four stewardesses. All meals were cooked on board and, in the days before trollies, drinks and food had to be carried to passengers on trays. Former stewardess Maureen Blaydon recalled that on London–San Francisco flights it was normal for two full meals and a hot snack to be served.

But if the early 707s lacked the ability to make non-stop Atlantic crossings, that was soon remedied by the bigger 707-320s, the first of which entered Pan Am service in July 1959. The airline also offered greater frequency than BOAC could manage with its Comets, so in the 1960s it was the 707s and DC-8s that were carrying the bulk of trans-Atlantic traffic. Until, that is, the advent of the wide-bodies.

The First Trans-Atlantic Jets

Jet-propelled aircraft had actually been flying across the Atlantic for ten years before the Comet. But in July 1948 the pilots of the single-seat Vampire fighters who flew from Scotland to Canada in several hops, the first jets to cross the Atlantic, had been bundled up in protective overalls and oxygen masks. They had certainly not been relaxing in air-conditioned and pressurised comfort.

The six twin-boom fighters of No. 54 Squadron RAF were escorted by a Mosquito reconnaissance aircraft. Three York transports of

24 Squadron carried ground crews and support equipment as well as three relief pilots. The jets were led by Wing Commander D.S. Wilson-MacDonald and they completed the flight from the UK to Canada in three legs. They left Stornoway in the Hebrides for Meek's Field, Iceland. They covered 662 miles (1,060km) in two hours forty-two minutes before flying on to Bluie West 1, Greenland (757 miles or 1,211km in two hours forty-one minutes) and then Goose Bay, Labrador (783 miles or 1,250km in two hours fifty-five minutes). Total flying time was eight hours eighteen minutes.

The aircraft cruised at heights between 25,000ft (7,260m) and 32,000ft (9,750m), far beyond anything the early pioneers could dream about. Air Ministry press officer Thomas Cochrane flew in one of the accompanying Yorks. At such altitudes, he observed, 'the jet airliners of tomorrow will possibly travel, and the experience gained by this military operation will not be without its value to civil operators of the future'.

The RAF pilots still reported encountering strong headwinds, which at times reached 207mph (330kph). At other times, Cochrane reported, conditions at several of the airfields were 'by no means ideal'. Even so, the formation 'left Iceland at lunchtime and were in Canada in time for supper'. Each of the nine 54 Squadron pilots flew two legs of the trans-Atlantic crossing. *Flight* reported that:

> At Goose Bay the squadron was given a great reception by the RCAF. Also there to meet them was Col Dave Shilling, one of the US Air Force's greatest fighter pilots, who was waiting to take off with 16 Shooting Stars on the reverse of the route the Vampires had just covered.

The first non-stop east–west crossing by a jet came less than three years later. On 21 February 1951 RAF English Electric Canberra B.2 bomber WD932 piloted by Squadron Leader A. Callard of the Aircraft and Armament Experimental Establishment flew from Aldergrove, Belfast, to Gander. Callard covered the 1,800-mile (3,300km) journey in four hours thirty-seven minutes. The bomber was being flown to the US to act as a pattern for the Martin B-57, the licence-built version of the Canberra.

In an echo of the wartime Atlantic ferry operation, more than 400 Canadian-built jet fighters were flown across the ocean between December 1952 and May 1954 for service in the RAF. As a stop-gap pending the arrival of British-designed swept-wing jets – notably the Hawker Hunter – the RAF ordered Canadair CL-13s, licence-built North American F-86E Sabres. The order was funded by the US

Mutual Defence Assistance Programme. After testing, all RAF Sabres were ferried to the UK under Operation Becher's Brook, named after the famous steeplechase jump on the Grand National racecourse at Aintree.

The British government initially ordered 370 aircraft for service in Germany with a further sixty for service with Fighter Command at home. Deliveries were undertaken by No. 1 (Long Range) Ferry Unit (later No. 147 Squadron), which ferried the aircraft in batches of fifty or more aircraft at a time in stages over the 3,100 miles (4,960km) from Cartierville via Goose Bay, Labrador, Greenland, Iceland and RAF Kinloss. They were generally flown across in groups of thirty divided into six flights of five aircraft. On arrival in the UK the Sabres were flown to No. 5 Maintenance Unit at RAF Kemble, Gloucestershire, where they were painted in the current RAF camouflage paint scheme of dark green and grey before being delivered to operational units.

When the RAF called for volunteers to fly the Sabres from Canada, Flight Lieutenant Alastair Aked, who was serving on a Meteor squadron, decided to add his name to the list. He was posted to RAF Abingdon for the Sabre conversion course, where he logged eight hours on the 'delightful' jet before joining the ferry unit. In February 1953 Aked boarded a Transport Command Hastings for the flight to St Hubert airport, Montreal.

There, he wrote, the ground school was 'intensive and very professional'. He recalled: 'The lovely shiny Sabres were being delivered from the Canadair factory at Dorval but our Canadian friends wouldn't let us touch them until we had completed their winter survival course.'

That meant lectures on all aspects of survival should they have to bail out into the frozen wilderness with temperatures that could reach minus 30°C. Between the pilot's seat and the parachute was a survival kit, which included an axe, radio, emergency rations and a folding rifle. To put theory into practice, the jet pilots were taken to a frozen lake near the River Saguenay. There they were spread out until each man was on his own to undertake the first survival task, building a shelter by cutting down trees and lighting a fire.

They were in the wilderness for two days. They returned to civilisation full of confidence that they could survive harsh conditions if they had to. The survival course was followed by advanced technical training on the Sabre, followed by instrument flying in Canadair Silver Stars, licence-built Lockheed T-33s. Only then were the pilots able to get their hands on the Sabres for fuel flow and endurance testing before enough aircraft were assembled for the first stage of the journey, from Goose Bay to Iceland.

They had been reassured that should anything go wrong en route the USAF had stationed several Grumman Albatross amphibians along the way to rescue downed pilots if need be. Weather information in those days before satellites was provided by fourteen ships stationed at points across the ocean between Canada and Greenland, between Greenland and Iceland and between Iceland and Scotland. There were also five radio beacons to aid navigation. These precautions were necessary because of the Sabre's short range and the lack of diversionary airfields.

The accompanying ground crews travelled by Hastings or RCAF North Stars. With hangarage only available at Goose Bay, the ground crews often had to work in difficult conditions to maintain the aircraft. Aked recalled that 147 Squadron suffered three fatalities during the twelve months of Operation Becher's Brook. Another pilot had to eject because of a spurious fuel warning. 'We were told,' Aked remembered, 'that if ever the red light came on we should eject within seconds since the aircraft would explode because it meant there was a fire near the fuel tanks.'

In this case the warning light came on over Scotland and, as the pilot descended under his parachute, he had to watch his stricken Sabre plunge into the Solway Firth, taking with it the 'goodies' he had bought in Canada that were then unavailable in Britain.

By that time 147 Squadron had around thirty pilots qualified on the Sabre. Aked recalled that the largest single delivery began on 20 April 1953 when a double shuttle of more than fifty aircraft was flown to Kinloss. It arrived on 1 May. After each leg the pilots were flown back to their departure airfield to collect another aircraft. When the Hunter replaced the Sabres the Canadian-built jets were returned to the USAF in 1956 through RAF maintenance units.

The Brabazon Committee

The Second World War was far from won in 1942 when the British government turned its attention to the post-war shape of civil aviation. That autumn it appointed the Transport Aircraft Committee to 'advise on the design and production of transport aircraft' and to 'prepare outline specifications of several aircraft types needed for post-war air transport'. A key consideration was ensuring the industry's post-war capacity was used productively during the transition from war to peacetime work.

Inevitably this group came to be known as the Brabazon Committee after its chairman, former Cabinet minister Lord Brabazon of Tara. Among its members were Sir Francis Shelmerdine, a former director general of civil aviation, and William (later Sir William) Hildred, director general of the International Air Transport Association.

Work started work in earnest in December and the committee's report, delivered to the War Cabinet in February 1943, contained a list of five aircraft types on which it believed preparatory work should begin at once, although no specific manufacturers were linked to the concepts at that stage.

To build on this work another committee was formed under Brabazon's leadership. Between 1943 and 1945 it issued a number of interim reports before producing a final one refining its previous conclusions. But implementation of the resulting recommendations was to fall short of the committee's hopes and ambitions. 'More than half the proposals fell by the wayside,' noted Sir Peter Masefield, journalist turned civil servant and airline boss.

Even so, some of the types emerging from the committee's deliberations – the initial list of five basic projects had risen to seven by 1944 – could certainly be considered successful. In addition to what would later become the Comet, the list included the Vickers Viscount, the world's first turbine-powered airliner, the Bristol Britannia and the more conventional Airspeed Ambassador. And, with 542 built, the de Havilland Dove feeder liner became one of Britain's best-selling post-war civil aircraft.

Professor Keith Hayward, research director of the Royal Aeronautical Society, has pointed out that Brabazon and his colleagues had to work under difficult conditions. Inevitably many of their assumptions and much of their evidence 'were derived from pre-war notions about air travel'. He noted: 'Accurate information about future trends, especially developments in the United States, was hard to obtain and the exigencies of war ruled out elaborate surveys and design studies.'

Hayward saw the committee's work as 'a credible, if not always accurate, prediction of future civil requirements'. In spite of its limitations, 'it formed the backbone of government policy towards post-war civil aircraft production'. What elevated the committee's deliberations was its appreciation of the opportunity offered by Britain's substantial lead in the field of jet engine design to overcome American superiority in conventional piston engine airliners.

Four of the Brabazon types called for either prop- or turbo-jet engines. Clearly the most visionary proposal was for the Brabazon IV. Made less than two years after Britain's first jet aircraft had flown, it called for a jet-propelled mail carrier with North Atlantic capability. In mid-1944 the committee issued an interim report that recognised trans-Atlantic range was probably expecting too much at that stage. Accordingly, it recommended a jet airliner that could be used on European and Empire routes with seating for fourteen passengers and the ability to cruise at 450mph (720kph). It might, though, be followed by one for Atlantic operations using turboprop or ducted fan power.

Clement Atlee's Labour administration inherited the work done by the Ministry of Aircraft Production to implement the Brabazon programme. The new government could hardly fail to be aware that aircraft manufacture had become one of the nation's most important industries and, apart from any other benefits, the production of airliners would help to maintain employment in a strategically and economically important sector. And if it turned in a profit then so much the better.

There was also pressure from the Ministry of Supply, which had absorbed MAP's functions, to advance the programme. According to Prof Hayward, 'officials were acutely conscious of the lead built up by the Americans and in their opinion unless the programme was rapidly and comprehensively implemented the gap would grow even wider'.

The Turbine Pioneers

Bristol Britannia 310 Specification

Crew	4 to 7
Passengers	139 in coach class
Length	124ft 3in (37.88m)
Wingspan	142ft 3in (43.36m)
Height	37ft 6in (11.43m)
Maximum take-off weight	185,000lb (84,000kg)
Power plant	four Bristol Proteus 765 turboprops each generating 4,450ehp (3,320kW) each
Cruising speed	357mph (575kph) at 22,000ft (6,700m)
Range	4,430 miles (7,129km)
Service ceiling	24,000ft (7,300m)

De Havilland Comet 4 Specification

Passengers	81
Wingspan	114ft 8in (35m)
Length	111ft 6in (33.9m)
Height	29ft 6in (8.99m)
Power plant	four Rolls-Royce Avon 524 turbo jets each generating 10,500lb (10,500kN)
All-up weight	162,000lb (73,483kg)
Cruising speed	503mph (810kph)
Cruising height	42,000ft (12,800m)
Maximum range	3,225 miles (5,190km)

Chapter Fifteen

Queen of the Skies

The arrival of the jets revolutionised trans-Atlantic air travel to an extent no one had foreseen.

Spurred on by customer demands for the speed and comfort offered by the jets, the airlines were desperate to replace their piston-engined aircraft. By the summer of 1960, no fewer than twelve airlines had launched jet services between Europe and North America. Most were using Boeing, but in April 1960 KLM Royal Dutch Airlines began North Atlantic services with the Douglas DC-8. It was quickly followed by SAS, Swissair, Trans-Canada Airlines and Alitalia.

In 1962 more than 2.3 million passengers flew the North Atlantic, five times more than a decade earlier. Pan American continued to be the market leader by a substantial margin but, in an indication of the customer appeal of the new turbine-powered Britannias and Comets, the British Overseas Airways Corporation had risen to second place in the table. By the mid-1960s trans-Atlantic traffic was growing at an annual rate of 20 per cent and set to double every five years.

Pan Am's dynamic president, Juan Trippe, boasted that his airline's introduction of the Boeing 707 in October 1958 had been the most important aviation development since Charles Lindbergh's flight to Paris thirty-one years earlier. 'In one fell swoop,' he declared, 'we have shrunken the earth.'

The big new jets were indeed redrawing the air map of the world. The major international airlines were experiencing a surge in growth dwarfing that of most other industries. They were also catering for a new group of passengers. Tourists were changing the basis of airline economics by creating a fresh area of demand – particularly across the Atlantic. Newly introduced economy fares had brought a surge in demand that only the jets, bigger and more productive than their piston-engined predecessors, could hope to meet.

It was clear the shipping lines could not. In 1957, even before the jets, more passengers crossed the Atlantic by air than by sea for the first time. The development of the jet airliners and the huge expansion that followed created a far bigger mass market. In the time it took the *Queen Mary* to cross the Atlantic one 707 could carry the same number of passengers over the same distance.

Yet, while the world was coming to terms with this revolution in travel, the aeronautical engineers were already busy crafting the next one. The first results would appear on Stand L25 at London's Heathrow airport bright and early on Thursday, 22 January 1970.

Yet the first prototype of what was to bring about this second travel revolution in little more than a decade had flown for the first time only eleven months earlier. And it was more than just the project's abbreviated timescale that was causing hearts to flutter at Everett Field, Washington, that Sunday morning in February 1969.

Most of the Boeing people present were well aware that the company's future was at stake. Joe Sutter certainly was. The company's chief finance officer kept reminding him that his engineering team was spending $5 million every day on the world's first wide-bodied airliner. Sutter was confident the aircraft would fly well, but later admitted to feeling apprehensive as he pondered one question above all: do we have a good airplane or not?

The new Boeing 747 was a monster compared with the 707 and Sutter wondered what test pilot Jack Waddell and co-pilot Brian Wygle would make of it. But in an interview with the author, the man described as the father of the 747 recalled: 'As soon as they began to fly it we could tell over the radio that they liked the way it handled.'

They certainly did. The 747 was surprisingly light on the controls and felt good. And when Waddell rolled it 30 degrees to the left and then to the right, that felt good, too. From the company's F-86 chase plane Paul Bennett was reporting that everything looked OK. The 747 was certainly behaving better than the 707 had at this stage of its career. However, Waddell decided to cut the flight short because of a problem with a section of wing flap.

Sutter didn't stop worrying: there was still the landing to come. Waddell had practised with a special rig comprising a mock-up of the 747's flight deck and its sixteen-wheel landing gear towed around by a truck. Known as 'Waddell's Wagon', it would later be used to give pilots experience in taxying such a large aircraft at Boeing's training school. The pilot sat in a mock-up of the 747 flight deck built atop three-storey-high stilts. The pilot learned how to manoeuvre from such a height by directing the truck driver below him by radio. But Sutter was still worried about how Waddell would cope with a real landing. 'I was watching the thing come in on approach and wondering if they'd have problems,' he said. 'My temperature must have gone up 100 degrees watching that landing!'

Waddell can hardly have found it encouraging when, just before the flight, Boeing's president Bill Allen told him: 'Jack, I hope you understand that the future of the company rides with you guys this morning.' Sutter thought Allen probably regretted saying it; even after he had landed Waddell was still fuming. Sutter recalled him saying: 'Boy, that was a hell of a thing for Allen to say just before I went up in that airplane.'

It was hardly surprising that the successful flight generated huge relief. Decades later Sutter struggled to find words adequate to describe it. 'Elation was too mild a term for what we all felt,' he said. Sutter's wife, Nancy, burst into tears when the aircraft surged into the air, prompting him to write: 'I saw then that that she'd gone through just as much hell as I, or for that matter any of the 4,500 people on my engineering team.'

Yet the first flight wasn't the programme's most vital moment. That probably came in March 1966 when a deputation from Pan Am visited Seattle to view the two mock-ups Boeing's engineers had created. On their decision rested the future shape of the 747. Would it be a double-decker or would it be what they had just invented: the two-aisle wide-body?

Sutter had been appointed to lead the 747 engineering team in August 1965. The project was fairly low-key because much effort was going into other projects that took priority. Boeing was bidding to build a giant heavy lifter for the USAF, but the service would ultimately favour the Lockheed offering that was to become the C-5A Galaxy. Some historians believe that Boeing's bid led directly to the 747, but this is not so. The design philosophy behind the 747 was to develop a completely new aircraft. As Sutter put it: 'The only thing the C-5 brought to the table was the high-bypass ratio engine.'

Boeing was also busy with a supersonic transport that was supposed to trump the Anglo–French Concorde. The 2707 was ultimately cancelled, but not before it had exerted an indirect influence on the 747, as Joe Sutter explained:

> That [the 2707] was a high-technology effort that demanded a lot of up-and-coming aerodynamicists, structures and power plant people. The group that I was able to put to together actually came from the 707, 727 and 737 programmes. Some of them were getting close to retirement but people didn't think the 747 demanded super high-technology brains. It was just a matter of putting a good airplane together so it was a very good, well-balanced team.

With demand for air travel surging ahead in the 1960s the airlines, many of which had previously harboured doubts about the 707's size, were now clamouring for something bigger. The idea for the 747 had emerged from discussions between Allen and Juan Trippe. The Pan Am president was seeking an aircraft with enough range to operate routes such as Paris–New York and Rome–Chicago. He also wanted it to carry a decent amount of freight.

Trippe's secretary, Kathleen Clair, recalled how the two men had decided to go ahead with the 747. At that pivotal moment in commercial aviation history they were standing outside Trippe's office in the Pan Am building in New York, at one time the world's biggest office block. 'Allen said: "If you buy it we'll build it." Clair recalled: 'And Mr Trippe said: "If you build it I'll buy it." And they shook hands, and that handshake was better than a written contract. So they went ahead and did it.'

Boeing had been thinking in terms of a 250- to 300-seater, something like the extended 707-800 that the company currently had under consideration. But Trippe made it clear he wanted something that would accommodate 350 to 400 passengers. And he was in a hurry for it: Trippe was demanding delivery by the end of 1969. That gave Sutter and his team just twenty-eight months to design from scratch the largest airliner the world had ever seen.

At that point it was assumed that the new aircraft would be a double-decker. It would simply be too long otherwise. But that wasn't how Sutter and his team saw it. When they listed all the disadvantages of the double-decker, particularly its lack of efficiency, it was clear they needed to look for another solution. When they took into consideration the 8ft × 8ft (0.28m × 0.28m) cargo containers that were being standardised within the industry, it seemed a good idea for them to be loaded side-by-side. That gave the aircraft a more suitable length combined with decent cargo-carrying ability. Within the passenger cabin that offered ten-abreast seating in the tourist section and six to eight abreast in first and business class. 'There was no Eureka moment,' Sutter recalled. 'When we listed all of the problems we solved by going to the wide single decker it was pretty obvious that was the way to go.'

Now Pan Am had to be convinced. This led to what has gone down in Boeing folklore as the 'clothesline presentation'. The man chosen to go to Pan Am's New York headquarters and break the news about Boeing's favoured layout was Milt Heineman, a man with years of experience in designing passenger accommodation. His main prop for the presentation to Trippe and his senior executives was a 20ft (6m) rope. This he stretched across the Pan Am boardroom to demonstrate how wide the aircraft would be.

But the sceptical airline executives were still unconvinced. They had to see the two 747 mock-ups at Seattle for themselves, which they did in March 1966. On their decision rested the future shape of the 747: single or double-deck. 'When Trippe saw the wide single deck I think that's when he decided it was the way to go,' Sutter recalled. That was confirmed when Pan Am engineering director John Borger invited his boss to view the mocked-up flight deck. 'Trippe couldn't care less,' Sutter recalled, 'but when he saw the dead space behind the flight deck, he said: "What's this being used for?" I think Borger really did the programme a great service by saying it could be used as a crew rest area. All Juan Trippe said was: "This will be reserved for passengers."

In 1966 Boeing's board made the decision to proceed with the programme. By now the design had been frozen around a huge main deck able to support seating for up to 500 passengers in a high-density, ten-abreast three-plus-four-plus-three layout, although 350 was judged to be a more likely number. The main passenger area extended right up to the aircraft's nose, where the first-class accommodation would be located.

As General Electric's CF6 had been selected to power the C-5A, Pratt and Whitney's robust but untried JT9D was well-placed to be chosen for the 747, which it duly was. The GE unit would, however, be offered as an alternative, as would the Rolls-Royce RB-211. Boeing was not only breaking entirely new ground with the 747 itself but it was also using a brand new engine. And, at the same time, the 747's size – its tail was tall as a six-storey building – required a new 200 million cu ft (5.6 million cu m) assembly plant at Everett, 30 miles (48km) from the company's base at Seattle. It would be the world's largest building by volume.

Even though the decision to proceed with the 747 had been taken in April 1966, it was to be another month before Pan Am actually put pen to paper. The 780-acre Everett site was acquired in June, by which time Boeing had a payroll of around 105,000. It also had $1 billion riding on the project.

But if the financial risk was high the engineering problems to be overcome were colossal. Trippe had also been pushing for the highest speed that could be achieved for the new aircraft. Boeing, with its vast experience gained from the 707 and military programmes, conducted studies using its own wind tunnel. These showed that Mach 0.85 would be the most efficient cruising speed, but Pan Am specified Mach 0.9. Accordingly, an advanced wing design was chosen with a 37-degree sweep-back and an array of new and sophisticated high-lift devices.

The landing gear had to be capable of bearing the weight of the 350-ton monster and the installation of the engines, mounted in pods big enough for a man to stand upright inside, posed further headaches. As the engineering team worked through these problems, moves were being made to cut the team by 1,000 workers. In a presentation to the company's top management, now facing intense pressure from the banks, Sutter not only had to defend his team, but ask for another 800 staff. 'I had to tell Allen that if we dropped the 1,000 the whole damn project would collapse,' Sutter recalled. 'I tell you, I never felt more lonely in my whole life.' His team wasn't trimmed, but he didn't get the additional 800 either.

Preparation for assembly began in January 1967 and was well under way by the end of the year. With the order book standing at 158 aircraft for twenty-six airlines, the first 747 was rolled out in September 1968. The first flight was originally scheduled for 17 December 1968, the sixty-fifth anniversary of the Wright Brothers' 1903 flight, but production issues and trouble with the engines delayed it until the following February. Flight testing was to reveal further issues. Among them was a propensity for the engines to surge and stall. Worst of all was the tendency for the high compressor fan blades to rub against their casings at the bottom of the engine.

Another serious problem was flutter or instability over Mach 0.86. At this stage the team was anxious to avoid a costly and time-consuming redesign. An improvised solution was to reduce the stiffness of the pylons supporting the engines where the oscillation was most marked. At the same time, the two outboard pylons were beefed up to enable them act as counterweights.

The second 747 was rolled out in February 1969 and flew for the first time in April, a month later than planned. Four more had come off the assembly line by the end of 1969. That June the fourth example, registered N731PA, was displayed at the Paris Air Show at Le Bourget, the airport where Charles Lindbergh had landed in 1927. The 747 arrived in Paris nine hours eighteen minutes after its departure from Seattle's Tacoma airport. On the return journey one engine had to be shut down after a power surge, but the aircraft landed safely at Boeing Field.

By now Pan Am was having second thoughts about the 747. The promised level of performance had not been reached and range was still seven per cent lower than expected. The airline threatened to withhold $4 million from the final payment on each aircraft until the problem had been rectified. Despite being heavily in debt, and with a recession now beginning to bite, Boeing held its ground and threatened to sell the first 747 to TWA. A compromise was reached under which $2 million would be withheld on each 747 until the engine problems had been resolved, with the balance being paid in instalments. This enabled the first 747, N733PA, to be delivered to Pan Am in December, followed a week later by a second example. Full type approval was granted by the FAA on the last day of 1969, even though the engine problems were some way from being resolved.

Even so, in just seven years Boeing had created the world's largest airliner and done so from a virtually clean sheet of paper. But it was also clear that the aircraft's sheer size posed a host of new problems for the industry. Crews would need to be instructed in handling it, while airlines would need new ground handling equipment, together with procedures for processing larger passenger loads. Furthermore, airports would have to expand to cope with the 747's sheer size.

The fact was that the new airliner was 50 per cent longer than a 707, had a wingspan 35 per cent bigger and a maximum take-off weight two and a half times greater. Another key difference was that the 747's door sill was nearly 6ft (1.8m) further from the ground than the 707's.

The need to accommodate such an aircraft caused consternation among the world's major airports; they had less than four years to prepare. 'There was panic – that's not too strong a word,' recalled John Mulkern, then a senior executive at Heathrow airport, in an interview with the author. London's premier gateway would be the first foreign airport to receive the 747 when it entered passenger service, which it was scheduled to do at around the turn of the decade.

At Heathrow the long-haul terminal, Terminal 3, which had been in operation only since 1961, required complete redevelopment at a cost of £13 million. A new arrivals building, based on a converted office block with 170,000 sq ft (15,790 sq m) of floor area, together with a T-shaped pier running from its south-western corner, was to connect both it and the separate departures building to special 747-sized parking stands.

To ease the strain on passengers using this extended facility, the 900ft-long Pier 7 was to be equipped with a moving walkway, the first travelator to be installed at a European airport. Seven new Jumbo-sized stands were to be constructed initially and this required the closure of a redundant cross-runway.

But as Pier 7 would not be ready until May 1970, Pier 5 had to be adapted to cope with the 747's greater door sill height as a temporary measure. Pier 7's gate rooms, air conditioned and big enough for 500 passengers, came into operation in phases during 1970 with a further three brought into operation in 1971. Runways had to be strengthened and taxiway exits made wider. The two road tunnels passing beneath the runways also had to be beefed up.

In what John Mulkern described as a 'miracle of adaptation', the airport had 'got a brand-new facility into operation within five months of the arrival of the first Jumbo jet'. The new wide-bodied aircraft was to have a profound impact on airport operations, transferring the main operational pressure point from runways to terminals. Pan Am had set a target time of one hour to process full passenger loads using a newly introduced computerised system.

By 1971 New York's Kennedy airport would have a new $50 million terminal, which was said to be six and a half times bigger than the existing Pan Am facility. There would be an automated $2.5 million baggage system in which bags would travel by conveyor system from the check-in desk down one floor to ground level, where they were to be sorted into up to eighty different categories for loading. The airline expected that half an hour after the simultaneous arrival of two 747s, the 700-plus passengers should all have cleared the formalities, claimed their baggage and be at the kerbside awaiting ground transportation to take them to their ultimate destinations.

Much thought also went into the design of the seats on Pan Am's 747s, right down to the smallest detail. The ashtrays in economy-class seats were to be big enough to handle fifteen half-smoked filter tip cigarettes plus wrapping from the package. In addition to cabin attendants, the airline planned to carry on early flights what it described as customer service representatives. Their function would be to deal with passenger enquiries or complaints that would otherwise delay the service given to other passengers. They were described as a cross between travel agency couriers and mobile complaints departments. Harold Graham, Pan Am's vice president for service, said they were being hired 'on the concept that they have got to work themselves out of a job in two years. If they are good enough to do the job they won't be satisfied by it'.

The airline also had to invest heavily in specialised ground handling equipment. The cost of the equipment needed to service one 747 was put at $776,000, elevating the total investment required by the fleet as a whole to $18 million. On the credit side, Graham pointed out that the new aircraft, with its high bypass ratio turbofan engines, would be less noisy than current jets. It would also use less runway because of the extra engine power.

To prepare pilots for the new aircraft, a former Strategic Air Command B-52 base at Roswell, New Mexico, was operated by Pan Am as a training facility. Most of the pilots who trained there had flown 707s and DC-8s for six to eight years and found that the 747 was easier to handle than either of the current jets. It did, however, require a higher level of concentration to fly it accurately. The ideal candidates, though, also had experience of the 727 and 737 airliners. Although smaller, these Boeing products had characteristics in common with the 747. By June 1970, when more than 200 Pan American pilots had attended the course, it was taking an average eight hours forty-three minutes to train them in handling the new airliner. In addition, Pan Am had trained a further seventy American Airlines pilots.

Pan Am's 747 operating crews comprised two pilots plus a pilot-trained flight engineer. Despite the height above ground level of the 747's cockpit, trainee pilots soon discovered that it was possible to make what was described as the softest landing of any commercial jet. Its two engine-out approach was also praised as the finest.

As the date for the inaugural passenger flight approached it became clear that the engine problems might not be resolved. On 12 January Pan Am had planned a proving flight from New York to London as part of a European tour. But, as *Flight* reported, 'engine troubles and bad weather resulted in the cancellation of the rest of the tour'. On the 15th, the First Lady, Mrs Pat Nixon, christened N733PA, the aircraft Pan Am had chosen to make the first commercial trans-Atlantic flight, *Clipper Young America*.

But there was still another hurdle to be surmounted – the FAA's mandatory evacuation demonstration. Held at Pan Am's Roswell training facility, the first two attempts were only partially successful. A lighting fault hampered the first and on the second two escape slides failed to deploy. Boeing now had six days to come up with a fix if Pan Am's inaugural flight to London was not to be postponed.

The manufacturer succeeded in this but Boeing and Pan Am executives were still chewing their fingernails. It was a bleak, bitterly cold evening at New York's Kennedy airport when the first fare-paying 747 passengers filed on board the aircraft that was to take them to London – or so they thought. They had reserved their seats on the inaugural flight two years earlier at a cost of $375 for a one-way first-class ticket and they joined the three flight crew and eighteen cabin attendants aboard *Clipper Young America*. Flight PA2 was under the command of Captain Bob Weeks, chief of the airline's Atlantic division, assisted by Captain John Nolan.

But even then several last-minute problems delayed the flight. *Time* magazine reported that:

> Pan American executives knew that they would run into all sorts of problems in getting their 747 Jumbo Jets through what airmen call a new 'plane's 'learning curve', and they tried to anticipate as many as possible. But a variety of major and minor difficulties, some of which could hardly have been anticipated, turned the 747's first commercial flight from New York to London into an alternately frustrating and funny experience for the 352 passengers.

For one thing, the main portside door in the forward economy class cabin refused to close. But far more serious was the engine problem that became apparent as the aircraft was taxying. The flight was already thirty minutes behind scheduled when the 747 was pushed back from the gate but now the number four engine started overheating, necessitating a return. Concern about internal engine damage meant that the back-up *Clipper Victor* had to be used. In a move that has confused historians ever since, it was temporarily renamed *Clipper Young America*. Seven years later, though, *Clipper Victor* would be involved in what is still the world's worst aircraft accident when it collided with another 747 on the ground at Tenerife.

The first 747 finally left Kennedy at 01:52 hours on 22 January, arriving at Heathrow at 14:02.

It had been a stuttering debut and it would be some while before things improved. The engine problems persisted before P&W finally solved them. The fouling of the turbine casing was traced to it becoming oval in shape due to the way the engine was attached to the pylon. The difficulty of fixing the problem was compounded by the rate at which Boeing was now turning out the new aircraft – one every three days. P&W was finding it difficult to keep up. Completed aircraft had to be parked to await the delivery of engines. Strategically placed concrete blocks stopped them tipping back on to their tails.

Boeing felt P&W should have done more preliminary research and testing. There were discussions about who was going to pay the extra cost of getting the engine to perform as it should. 'They finally reached a conclusion and Pratt began working real hard to straighten out the issues,' Sutter recalled.

Nevertheless, these problems and their impact on reliability tarnished the 747's image during its early days of service. In a consumer report published six months after the aircraft's introduction to service, *Flight* observed: 'From the advertisements you might think that the 747 is a new dimension in travel. It is, but judging by the criticisms which have been made of the first few months of operation, much needs to be done before the entire extent of the "new generation" is exploited fully.' A reporter who took a trans-Atlantic trip with TWA experienced delays and cancellations. He wrote: 'More attention might be paid to passenger convenience at the expense of a little of the quite understandable competitive advertising.'

Pan Am found that the biggest source of complaints was the in-flight entertainment system, even though it was the most modern available. 'The system is very delicate,' noted Harold Graham. 'It either works or it doesn't.' A more fundamental problem was one that would strike a chord with later travellers: the availability of toilets towards the end of a long flight. The problem, Graham acknowledged, was that women liked to renew their make-up before landing. 'Each person can block a cubicle for up to 15 min,' he said, but he was aware that adding special cubicles for this purpose would mean fewer seats.

In terms of 747 deliveries, TWA was just behind Pan Am. Its aircraft were first operated on the New York–Los Angeles route. The first overseas airline to operate the 747 was Lufthansa, which began Frankfurt–New York operations in April. BOAC took delivery of its first example that same month, but an industrial dispute with pilots over the pay they should receive for flying the 747 delayed operations for a year. By the end of 1970 a dozen airlines were operating the 747 on a variety of long-haul and trans-oceanic routes.

In January 1971 Boeing delivered its 100th 747. The early troubles were now largely behind it as the Jumbo Jet became a familiar sight at the world's major airports. A total of 167 of the original -100 version was sold. Next came the improved -200 and the -300, the latter of which introduced the stretched upper deck. But the one produced in the largest numbers was the -400, which also featured winglets and raked wing tips. This increased wingspan to an extent that required some airports to expand again to maintain clearances.

As the 747 approaches its half-century a dwindling number remain in service. Of the type's major users, United Airlines retired its last example in 2017 and British Airways, which has the biggest fleet of 747-400s, plans to phase them out in 2024. Meantime, the modernised and improved 747-8 remains in production, although it has been selling in small numbers compared to big twins such as the Boeing 777 and 787 Dreamliner. Nevertheless, in 2014 Boeing delivered the 1,500th 747 to come off the production line to Lufthansa, making it the first wide-body airliner to reach this production milestone.

Over the last half century, the aircraft widely known as the Queen of the Skies, has carried more than 3.5 billion people, equivalent to half of the world's population, and it has been used by operators in eighty-nine countries to move people and carry freight.

Few aircraft have caught the public imagination in the way the Boeing 747 has. And few airliners have changed the way we fly as profoundly as the first of the wide-body jets.

Boeing 747-100/200 Specification

Flight crew	3
Passengers	374 to 490
Length	231ft 10in (70.66m)
Span	195ft 8in (59.6m)
Height	63ft 5in (19.3m)
Gross weight	735,000lb (330,000kg)
Power plant	four Pratt & Whitney JT9D-3 turbofan engines each generating 43,000lb (193kN) thrust (alternatively, four General Electric CF6 or four Rolls-Royce RB211-524)
Cruising speed	640mph (1,024kph)
Range	6,000 miles (9,600km)
Service ceiling	45,000ft (20,400m)

Boeing 747-400 Specification

Flight crew	2
Passengers	416 (3-class), 524 (2-class)
Length	231ft 10in (70.6m)
Wingspan	211ft 5in (64.4m)
Height	63ft 8in (19.4m)
Max take-off weight	875,000lb (396,890kg)

Power plant	four Pratt and Whitney 4056 turbofans each generating 63,300lb (2,823kN) thrust; four General Electric CF6-80C2B1F each generating 62,100lb (276kN) thrust; four Rolls-Royce RB211-524-11 each generating 59,500lb (265kN) thrust
Cruising speed	567mph (912kph)
Range with max payload	7,260 miles (13,450km)
Service ceiling	45,100ft (13,747m)

Chapter Sixteen

Revolution in the Air

Some people had queued all night to get tickets. Freddie Laker, though, had been waiting six years to launch his Skytrain cut-price, no-frills trans-Atlantic operation and he was determined to savour every moment. A quarter of a century later he admitted: 'I still get emotional about it.'

The queues were understandable: you could fly on Skytrain from London to New York for the price of a decent suit or a meal in a good West End restaurant.

During the 1960s growing incomes had raised aspirations and stimulated a demand for air travel among people who could not previously have contemplated flying across the Atlantic. As the smaller independent airlines sought ways around the rules put in place by governments to protect their national airlines, tour operators and airlines were offering inclusive tour holidays. This meant that flights and accommodation were packaged together and sold at prices the market would bear rather than what was decreed by governments.

Initially, most holidaymakers were content to fly to Mediterranean resorts and soak up the sun there, but it was not long before the more adventurous were demanding correspondingly low fares between Europe and North America. The travel industry sought to oblige them and in doing so created some anomalous situations.

Ironically, this move was facilitated by the International Air Transport Association, the scheduled airlines' trade association, which worked hand in glove with governments to fix international air fares. An obscure IATA rule permitted airlines to carry members of clubs or associations formed for purposes other than travel at fares below IATA's officially agreed minima. It also stipulated that tickets had to be booked at least three months in advance by bona fide, fully paid-up members of an officially recognised organisation.

Among the first airlines to spot the potential of the so-called 'affinity group' charter rule was the newly formed Caledonian Airways. Founded by former airline pilot Adam Thomson and airline executive John de la Haye, it sought to exploit Caledonian's Scottish heritage and the strong ethnic ties between the big expatriate communities in North America and their home country.

Before starting operations, however, Caledonian needed permission from the authorities in the US and UK and it was clear this would not be easy. The new airline faced strong opposition from the established carriers, particularly Pan American and BOAC. They feared that granting Caledonian a licence to operate its proposed charter services would open the floodgates to low-cost travel over the North Atlantic. It was not until May 1963 that the Civil Aeronautics Board (CAB) finally granted Caledonian its foreign carrier's permit. When it was signed a month later by President John F. Kennedy, it represented a breakthrough not just for Caledonian, but also for other airlines keen to operate trans-Atlantic charters.

By 1970 Caledonian was flying 800,000 charter passengers a year and had become the dominant UK player in the affinity group market, which was still relatively untapped. One of the biggest of these groups was the Paisley Buddies, many of whose members were mothers and grandmothers from Glasgow who had never flown before. They were totally genuine, as were bodies such as the Midlands Dahlia Society and many others, but there were doubts about those such as the Rutland Cricket Club and the Trowbridge Caged Birds Society. In 1970, 8 million passengers flew across the Atlantic and around 1.4 million were flying as members of affinity groups. It was estimated that as many as half were not actually members of bona fide organisations.

It was becoming obvious that fake 'organisations' were being created, not to grow dahlias or breed caged birds, but simply to fill a Boeing 707 for a trans-Atlantic charter flight. Bogus membership cards were issued, sometimes on the day of travel with membership back-dated to suit the rules. 'Specialist' travel agencies had suddenly sprung up on both sides of the Atlantic to cash in on this cheap flights bonanza.

By the early 1970s governments on both sides of the Atlantic were becoming aware of the scam and began policing departure airports. Government inspectors would swoop just before a flight was due to depart and demand proof of club membership. Passengers unable to convince the inspectors of their credentials were turned off the aircraft. It was in the nature of such flights that this often happened late at night or in the small hours of the morning. The disappointed mums and grannies who had been saving up for the opportunity to be reunited with long-lost relatives in Canada or the USA provided excellent copy for the news media.

Laker Airways had come late to the trans-Atlantic charter market, having acquired two Boeing 707s in 1969. Its founder, 'the rumbustious, buccaneering' Freddie Laker tried to impose his own rules by asking his passengers to swear on the Bible before boarding that they had been members of their affinity group for a year and that they had not joined just to travel.

But it soon became obvious to Laker that his faith in ordinary men and women not telling lies just to get a cheap flight had been misplaced. In March 1971, thirty-eight passengers had to be turfed off a 707 about to leave Gatwick for New York following a warning from the Department of Trade and Industry. Two months later, forty-six passengers were revealed to be falsely posing as members of the US Left Hand Club. The aircraft was delayed for three hours thirty minutes as inspectors interrogated the would-be passengers, in a raid filmed by Independent Television News.

This was the last straw for Laker. He would later maintain that it provided the motivation for the launch of his cut-price Skytrain service. In fact, he had already registered the name a year earlier. In June 1971, he told his staff: 'What we want is something simple like a train.' His idea for a cheap walk-on scheduled service between London and New York might have been popular with potential passengers, but the authorities on both sides of the Atlantic would try hard to stifle it. Laker's persistence, coupled with his growing status as something of a folk-hero and champion of the 'forgotten man', eventually got Skytrain off the ground. But it would take six years. Even Laker could not have imagined how difficult it would be.

He was, though, in no doubt about the implications of what he was proposing. In June 1971 he told a press conference that the single summer fare of £37.50 he then proposed to charge for flights to New York represented a direct challenge to the UK and US governments to liberalise air travel and reduce fares.

Meanwhile, the charter rules were getting tougher. The US CAB revealed that it had compiled a list of 217 charter flights by Laker Airways that it claimed had included bogus affinity groups. Also affected was another British operator, Donaldson International. The CAB required both British carriers to submit information twenty-five days before each flight to prove that they were abiding by the regulations. The impact on Laker was profound. The number of trans-Atlantic charters his airline was able to operate was slashed from around 150 in 1971 to about a dozen the following year.

But then the authorities and the travel trade reached a compromise. The affinity group rules would be replaced by the much simpler Advanced Booking Charters, which required a four-week advance booking period. This was later reduced to two weeks. Laker refused to see ABCs as an alternative to Skytrain. He argued that abuses would creep in, as had happened with the affinity groups. Charter and scheduled operations were different and, besides, Skytrain would appeal to a wholly new market.

He had spotted an opportunity to acquire what seemed to be the ideal equipment for it. With 345 seats compared with the 707's 150, the wide-body McDonnell Douglas DC-10 seemed likely to offer favourable operating economics. And the break-even load factor of 52 per cent compared with 63 per cent for the 707 seemed almost too good to be true. The biggest problem seemed to be financing the purchase of two of the DC-10-10s that were now available. True to form, Laker would pull off a good deal with their owners, the Japanese conglomerate Mitsui.

As it happened, one of these newly acquired tri-jets operated the first ABC flight when, in March 1973 it took off from Manchester airport with 250 passengers bound for Toronto. Laker now seemed to be on a roll. After a three-day hearing that autumn, the Civil Aviation Authority decided to grant Laker's Skytrain application. There were conditions – off-season capacity would be limited to the capacity of a 707 and it would have to be based at Stansted airport in Essex – but Laker now had a ten-year operating licence.

Under the terms of the US–UK air services accord, the Bermuda Agreement, the US authorities were obliged to endorse Laker's designation and issue a foreign carrier's permit. Instead they dragged their feet, no doubt keen to preserve the position of Pan Am and TWA. That was how matters stood in March 1974 when there was a change of government in the UK. The incoming Labour administration was even more keen than its Conservative predecessor to preserve British Airways' position. The view of its back-bench Labour MPs was that Laker was 'a johnny come lately airline' out to cream off part of the trans-Atlantic market at the nationalised carrier's expense.

Against this background Laker was becoming increasingly suspicious of the new government's true intentions. Trade secretary Peter Shore and junior minister Stanley Clinton Davis assured him they were continuing to press the US government to issue Laker's permit. But Laker was convinced that they were really working to thwart his plans. Matters came to a head when he accused Shore of acting illegally.

The government's subsequent response to the capacity crisis that was now developing, particularly in the North Atlantic market following the quadrupling of oil prices in late 1973, was to create a new civil aviation policy. Naturally it favoured the state-owned BA, although it did recognise the need for giving leading independent airline British Caledonian room to continue operating. Laker got nothing.

He was furious, appearing at an international conference to denounce the 'bums and gangsters' at the Department of Trade. More than thirty years later Lord Clinton Davis was unrepentant. He told the author: 'BA had 94 per cent of the UK aviation market with British Caledonian having 6 per cent. That seemed to be working very well. Peter Shore and I thought Skytrain wasn't a runner. We had no faith in Laker Airways itself. Running a charter operation is quite different from a scheduled airline.'

When Shore called a press conference at the trade department's London headquarters to announce the withdrawal of Laker's designation it was gate-crashed by a furious Freddie Laker. As the minister beat a hasty retreat, the journalists brushed aside the officials' protests and urged Laker to take the floor.

Laker then took his case to court. He lost but appealed and the result was a ruling in his favour. Master of the Rolls Lord Denning found that in seeking to block Laker the government was acting illegally. By then Shore had been succeeded by the more moderate Edmund Dell, who decided against an appeal to the House of Lords. 'He was in favour of the Laker experiment,' Clinton Davis recalled. 'It wasn't a resignation matter but I thought he was wrong and told him so. We had many discussions on the issue. They were very polite and I had no hesitation in saying where I stood, but I was only a junior minister and I had a lot on my plate.'

The change of heart meant that the civil servants Laker had previously denounced were now working to support his Skytrain service. Laker would later say that Dell had assured him: 'We're behind you 100 per cent.' Indeed, the way the US authorities had earlier dragged their feet on the issue was one of the reasons why, in 1976, the UK tore up the Bermuda Agreement and called for its renegotiation. Patrick Shovelton, the official who led the British side, recalled:

> In the event we grafted Laker nicely into Bermuda II negotiations. In return for getting Freddie into New York we allowed Pan Am and TWA to continue into Heathrow.

President Jimmy Carter gave his approval and now it was BCal's turn to be furious. It had been implacably opposed to Skytrain, fearing that the New York route could not support two British independent airlines. It was right. In the event only BA and Laker were to offer scheduled services on the route. Laker Airways was subsequently allowed to add Los Angeles to its schedule.

The first Skytrain service left Gatwick at 16:28 hours on 26 September 1977. For Laker it was a moving occasion. A total of 234 passengers from twenty-two different countries had been queuing for the £59 one-way tickets since the previous day. The airline had not been able to sell them until noon on the day of departure. First in line to be greeted by a beaming Freddie Laker was Anne Campbell, a fashion buyer from south London. She said she had been waiting for fifty-five hours.

Laker spent much of the journey to New York on the flight deck of the DC-10 *Eastern Belle* (G-AZZC) with Captain Basil Bradshaw and first officer Michael Willett. Sitting in the supernumerary crew position, the airline boss carried out a series of media interviews. 'It was quite funny,' Willett later told the author. 'As we got airborne he was giving interviews over VHF radio until we got to Ireland and then over the HF [high frequency] radio after that. He also did TV interviews.'

Towards the end of the flight, bad weather made it seem that it would have to be diverted. Looking tired and haggard, Laker appeared from the flight deck to announce that, instead of a triumphal arrival at the Big Apple, the flight might be diverted to Boston. It would have been a public relations disaster for the pioneering cut-price, no-frills operation. 'It wouldn't have been very good on an inaugural,' acknowledges Michael Willett. 'Everybody ahead of us diverted but when we got to New York the weather was just OK.' *Eastern Belle* landed after a flight of eight hours twenty minutes.

As Skytrain appeared to be going from strength to strength – Miami and Tampa were later added as destinations – Laker received his reward from a grateful nation. He was knighted in 1978 in the final *volte face* by a government he had battled against so long and so bitterly. But the opposition was fighting back. In 1978 the other trans-Atlantic carriers responded by offering cut-price 'stand-by' tickets. Initially, they were sold at Heathrow, incurring the wrath of the British Airports Authority, which took BA, Pan Am and TWA to court to prevent this happening. As a result, the queues were transferred from Heathrow to central London.

Despite Skytrain's popularity, Laker Airways was on shaky ground. Another recession and the grounding of DC-10 aircraft following an accident at Chicago's O'Hare airport did not help. But grandiose expansion plans meant the airline was over-committed and not in good shape to survive the latest economic downturn. And it was little consolation to Laker that a law suit against rival carriers who were found to have colluded to weaken the Skytrain operation was successful. In any case, it was too late: on 5 February 1982 Laker finally had to accept that his airline was finished. But the genie was out of the bottle. Skytrain had created expectations and things would never be the same again.

It was not the only new operation to arise from the renegotiated US–UK air services agreement, better known as Bermuda II. Yet in the summer of 1976 such benefits had seemed unlikely to the journalists being briefed by trade department officials on the UK government's decision to denounce the original Bermuda Agreement. The experienced air correspondents were incredulous that the UK could hope to win a better deal from the world's most powerful aviation nation. Edmund Dell told MPs that what Britain was after was a bigger share of North Atlantic traffic for British airlines and greater access to the US market.

This caused no surprise in Washington. Despite the long-standing friendship between the US and the UK, the exchange of air services between the two nations had been a source of constant friction since the original treaty was sealed. According to Prof Martin Staniland of the University of Pittsburgh, the British saw Bermuda I as wasteful and unremunerative. 'The waste arose from the lack of restrictions on capacity embodied in the 1946 agreement: American carriers had reacted to the [oil price] crisis by forcing down fares and increasing trans-Atlantic services at the expense of British Airways' revenue.'

Official US statistics showed that in the early 1970s American carriers were earning £350 million annually on the trans-Atlantic route, while their British counterparts were making little more than a third of that, £120 million. Staniland was also convinced that the British were 'intent on controlling capacity more stringently so as to increase BA's earnings and ease Britain's balance of payments problems'.

But the UK's current economic difficulties represented just one factor. The Bermuda II negotiations were conducted against the background of US opposition to Concorde commercial flights to New York and the challenge to the British policy of single-designation on trans-Atlantic routes posed by Laker's Skytrain. Moreover, the Labour government had suspended BCal's rights to operate to Houston and Atlanta. This, observed *Flight*, had not made the British position stronger. And the fact that Laker had just secured a High Court judgement against the government, 'makes HMG appear even more muddle-headed', in the journal's opinion.

The British negotiators proved unexpectedly tough. It took nearly a year to conclude a deal and then only after the intervention of President Carter and Prime Minister Callaghan. Until then both sides had been making contingency plans to operate trans-Atlantic services in the absence of a formal agreement. In the event, the new agreement was signed ten minutes after the old one expired.

One of the many complicating factors had been the position of British Caledonian. This had led to friction, albeit mostly on the British side. Although Caledonian's acquisition of British United in 1970 had been seen by the Heath government as a move towards the creation of a 'second force' airline, the current Labour administration was not so favourably disposed towards Adam Thomson and his airline. Accordingly, Thomson was deeply suspicious of the government's intentions towards BCal.

Under the terms of the treaty, double designation was ended on all routes except London–New York, on which Laker was to be the second British carrier, and London–Los Angeles, with BCal partnering BA. The US carriers' fifth freedom rights over London to four European cities was also to end and capacity would be subject to prior annual approval. British carriers were given rights to operate to six additional US cities.

Even so, Thomson felt his suspicions had been justified and that BCal had been let down, particularly in losing his New York designation to Laker. But there was some consolation that BCal could operate on the Houston and Atlanta routes, even though the latter would be in competition with Delta. And there was provision for Braniff to operate between Dallas/Fort Worth and London.

Yet, while Thomson grumbled about what he saw as favouritism for BA and the government covering its embarrassment over the Laker court case resulting in 'a bodged treaty and financial disadvantage', the Americans thought the accord favoured the Brits. Prof Staniland saw Bermuda II as 'distinctly and perhaps surprisingly advantageous to the UK'. Alfred E Kahn, chairman of the US Civil Aeronautics Board and a leading proponent of deregulation, would later say that Bermuda II was based on terms that were 'so contrary to our fundamental competitive principles that many of our airlines were astounded'.

The British government hailed the agreement as a victory worth 'tens of millions of pounds' for the nation. The Conservative opposition, though, supported Thomson's view. BCal's response was to advance the launch of its Houston service by six months to October 1977, the month after Laker launched Skytrain.

The treaty also enabled the British to establish the principle that Heathrow would be barred to airlines new to London. This meant that services such as Delta's Atlanta–London and Braniff's Dallas/Fort Worth–London operations, which began in the spring of 1978, had to use Gatwick. This triggered a last-minute diplomatic spat.

This principle was reinforced in 1980 when, in an apparently minor revision to the treaty, the UK agreed to double-designation on the London–Boston route. In return the US agreed that only two US carriers – then Pan Am and TWA – should have access to Heathrow. When they were subsequently forced to withdraw from trans-Atlantic operations their places were taken by United Airlines and American Airlines. The price, however, was expanded rights for BA and Virgin Atlantic.

In 1995 an annex to Bermuda II lifted the restrictions on scheduled US–UK services using only Heathrow and Gatwick. The result was that a new generation of all-business class carriers including Eos, MaxJet and Silverjet were able to operate from Luton and Stansted, although economic factors were to make these services relatively short-lived. Continental Airways also took advantage of the new provision by operating to the USA from UK regional airports including Bristol, Birmingham, Manchester and Glasgow.

But the key outcome of Bermuda II was that for thirty years Britain was able to resist the pressures that led other countries to sign open skies agreements with the US. That came to an end in 2007 when a new open skies accord was agreed by the US and the European Union. Among other things, it opened up Heathrow to US carriers from March 2008. But at a price: slots at the world's busiest international airport were no easier to come by. Continental Airways, which had hitherto operated at Gatwick, was rumoured to have paid $30 million for a pair of slots.

Despite Thomson's misgivings, BCal was able to make the London–Houston work to its advantage. Under Bermuda II it had three years to operate it solo before being joined by American Airlines. This was the time when North Sea oil and gas exploration was in its infancy and, by linking the Houston operation with its Aberdeen and West African services, BCal was able to claim it was the oilman's airline.

Martin Horseman, editor of *Aircraft Illustrated*, travelled on BCal's Flight BR246 from Houston to Gatwick aboard DC-10-30 G-BEBL *Sir Alexander Fleming* and wrote a detailed account of the journey. He reported that the airline had been conducting an extensive advertising campaign in the USA that emphasised its high level of on-board passenger service, but also attempted to promote Houston as a hub for trans-Atlantic traffic. As a result, passengers were joining Flight BR246 from origins as far afield as Denver, Tulsa and Oklahoma. The oil industry links, Horseman noted, 'have triggered significant traffic for BCal'.

The inbound flight, 245, from Gatwick, touched down at 15:55 hours one afternoon in 1979. It would be turned around ready for its early evening return flight with 239 passengers and crew on board. Captain Stuart Calder and first officer Bill Newman had filed a flight plan that called for Bravo Lima to route from Houston to the Newfoundland coast via the designated waypoints of Shreveport, Memphis, Nashville, Charleston, Philipsburg, Hancock, Albany, Kennebunk, Saint John, Charlottetown, Channel Head and Gander.

Exiting the Gander Flight Information Region, the DC-10 would be heading across the expanses of the North Atlantic, first into the Gander Oceanic Area and then to Shanwick OCA where, due to the earth's curvature, radar coverage is limited. In this area the aircraft would fly on one of many pre-determined tracks, its progress monitored by regular high-frequency radio reports.

As expected, the flight was uneventful. After overflying Ireland, via Shannon and Slaney, the DC-10 passed into the London Flight Information Region for the run to Gatwick along airway Upper Green 1 via Strumble, Brecon and Lyneham to Woodley and along Upper Amber 1 to Midhurst. Flight time, Horseman told his readers, had been eight hours fifty-three minutes, one minute earlier than the computer forecast.

Meanwhile, deregulation was gaining momentum. The system of rigid regulation of air fares and routes that had existed in the USA since the formation of the CAB in 1938 came under pressure in the 1970s. The effects of the oil crisis and technological advances brought by wide-body airliners were the key factors. The deregulation movement found its focus in a series of senatorial hearings into airline deregulation in 1975 presided over by Senator Edward Kennedy.

The hearings also showed that there was widespread cross-party support and in February 1978 Sen Howard Cannon of Nevada introduced the deregulation bill. It was signed into law by President Carter on 24 October 1978. The Airline Deregulation Act's main provisions included the encouragement of competition, enabling new airlines to enter the market and the progressive elimination of the CAB and its regulatory powers.

It brought profound changes to the industry. As the journal *Aviation Week* reported in a retrospective piece published in 2015:

> Before deregulation airlines competed on service alone, as fares were regulated by the government. Many remember this as the 'golden age of aviation' when stewardesses – as flight attendants were then known – carved chateaubriand on rolling silver carts and airlines put piano lounges in the upper decks of their Boeing 747s. Passengers dressed up to board flights and flying was glamorous and exciting – and mainly for the rich.

Another result was the removal from the market of some of its best-known and long-established names. Deregulation left international carriers including Pan Am, TWA and Braniff without robust domestic feeder networks, yet allowed domestic giants such as Delta Air Lines to apply for international routes.

Miami-based Air Florida, founded as a domestic airline in 1972, was soon operating international services. In 1981 it was designated as the second airline on the Miami–London route in competition with Laker's Skytrain. It began operations that April. Using DC-10-30 aircraft, the carrier offered a bookable fare that was £2 lower than Skytrain, although Laker claimed its standard walk-on flight was still cheaper.

By November, Air Florida was claiming that it had cornered 30 per cent of the London–Miami route in just seven months and was achieving an average economy-class load factor of 90 per cent. The airline expanded rapidly and was soon operating trans-Atlantic services to Amsterdam, Brussels, Dusseldorf, Frankfurt, Madrid, Shannon and Zurich as well as Gatwick. However, a serious accident in 1982, combined with a high level of debt, forced the company to seek bankruptcy protection in July 1984. Its assets were subsequently acquired by Midway Airlines.

Deregulation resulted in the rise of a new kind of airline such as Southwest, which had previously been confined to Texas by CAB rules. Airlines were allowed to innovate new business models. One of them was People Express, which, for a while, democratised air travel, including trans-Atlantic travel, in the way that Laker had tried to do with Skytrain.

In fact, its founder, Donald Burr, was already something of a Freddie Laker disciple, but he believed fares could be even lower and followed this principle with almost revolutionary zeal. He had been chief operating officer and then president of Frank Lorenzo's Texas International before leaving to pursue ideas to which the Deregulation Act had recently given fresh scope.

Burr formed People Express in 1980. He believed in cutting costs to the bone. To achieve a low cost base he put into effect a management style that could be described as revolutionary. He believed that air crews should be hired at a minimum salary and perform duties additional to those involved in flying the aircraft. This meant that pilots could find themselves working as baggage handlers.

Newark was selected as the airline's main base. Initially People Express operated domestic flights but on 26 May 1983 it began non-stop services between Newark and Gatwick using a leased Boeing 747-200 with 390 economy-class and forty premium-class seats. One-way fares started at $149 and the service was an instant success. The response was overwhelming. Within twenty-four hours every flight was booked solidly for the next two months. Passengers were able to bring one carry-on bag free, but were charged $3 for each checked bag, making it the first US airline to do so. If customers wanted food or beverages they paid extra: a can of soda cost 50c, beer 1$, honey roasted peanuts and brownies were priced at 50c and a 'snak-pack' comprising an assortment of cheese, biscuits and salami cost $2.

In 1985 flights to Brussels were offered and for a month were discounted to $99. By that time People Express was flying to forty-one cities worldwide with a fleet of seventy-two aircraft and 4,000 employees. But the same year the airline announced it was losing money partly due to an ill-judged corporate acquisition. Debt pressure obliged the airline to change its approach and in 1986 it shifted its focus to high-yield passengers by offering first-class accommodation in leather seats with meals served on fine china and crystal.

But People Express seemed to have lost its way and Burr's unconventional management style created a situation in which few people stayed in their jobs long enough to become familiar with their duties. In September 1986 People Express was taken over by Texas Air Corporation.

Yet People Express left a remarkable legacy. Captain Lynn Rippelmeyer had become the first female to fly a Boeing 747 with Seaboard World Airways in 1980. She joined People Express the following year. On 18 July 1984 Rippelmeyer became the first female 747 captain when she commanded a trans-Atlantic flight from Newark to Gatwick. In Britain she was widely acclaimed for this feat, being named Woman of the Year and receiving congratulations from Princess Anne.

McDonnell Douglas DC-10

The McDonnell Douglas DC-10 was the second of the first generation of wide-body airliners to appear and, although produced in larger numbers than the rival Lockheed TriStar, it was never able to approach Boeing's 747 output.

The aircraft arose out of design studies by Douglas for a higher capacity successor to the narrow-body DC-8 long-haul airliner. It would be smaller than the projected 747, but have a similar passenger appeal and economy of operation. Early planning went ahead against the background of Douglas's financial troubles, but these were resolved by the merger with McDonnell in 1967. Although the original studies called for two engines, three were eventually chosen to avoid problems at hot and high airports. American Airlines ordered twenty-five with options on a further twenty-five for its domestic routes. They would be powered by a modified version of the General Electric CF6 selected for the Lockheed C-5 Galaxy.

Competition with the TriStar meant that follow-up orders were slow in arriving, but in April 1968 United Airlines signed up for thirty plus options on a further thirty. Northwest followed soon afterwards, although it specified Pratt &Whitney JT9D-15s for longer-range operations.

It was, however, the DC-10-10, intended for domestic routes up to 3,500 miles (5,632km), that appeared first and it made its maiden flight in August 1970. It would be capable of carrying 255 to 270 passengers between Los Angeles and Boston, and American began operations with it in August 1971. United followed nine days later, launching its aircraft on the San Francisco–Washington route.

The first long-range variant, the -10-20 – later re-designated -10-40 – flew on P&W power, but the definitive long-range DC-10-30 used CF6s. This was the version favoured by European airlines including KLM, Swissair, Alitalia, SAS and British Caledonian.

Eventually forty-eight airlines bought 386 DC-10s, but its true potential was never realised due to a series of high-profile accidents and consequent negative publicity combined with sharing the market with the rival Lockheed product. DC-10s were modified as passenger/cargo convertible aircraft and sixty KC-10 aerial tankers were delivered to the US Air Force.

The last DC-10 was delivered in 1990, by which time the superficially similar but vastly improved and modernised MD-11 had been launched. British Caledonian was the initial customer, but the order was cancelled following the airline's acquisition by British Airways. By the end of 1986 the order book was standing at ninety-two from twelve customers. Production ended in 2000, by which time McDonnell Douglas had been merged with Boeing.

Most of the airlines that ordered MD-11s for long-haul passenger operations had replaced them with Airbus A330s, A340s and Boeing 777s by the end of 2004. Some were converted into freighters. A total of 200 MD-11s was produced.

McDonnell Douglas DC-10-30 Specification

Flight crew	3
Passengers	up to 380
Length	182ft 3in (55.5m)
Wingspan	165ft 4in (50.4m)
Height	57ft 6in (17.7m)
Maximum take-off weight	580,000lb (263,100kg)
Power plant	General Electric CF6-50C turbofans each offering 51,000lb (227Kn) thrust
Cruising speed	564mph (908kph) at 30,000ft (9,150km)
Range	4,600 miles (7,410km)
Service ceiling	33,400ft (10,180m)

Chapter Seventeen

New York in Three Hours

Brian Walpole was facing one of the sternest tests of his life. An experienced airline pilot and head of the British Airways Concorde fleet, Captain Walpole was on the flight deck of the first of the Anglo–French supersonic airliners to land at New York's John F. Kennedy airport.

As the city had tried hard to ban the aircraft from using the airport on environmental grounds, Walpole was expecting a particularly rough reception from the local news media. It was 19 October 1977 and Concorde 210 (F-WTSB) had arrived to conduct a series of proving flights that would test noise abatement procedures ahead of the start of scheduled operations by British Airways and Air France.

Aerospatiale test pilot Jean Franchi was in command and planned to take as many of the questions as he could, but Walpole still expected to be in the firing line. 'I went down to Toulouse,' he told the author, 'and we had a comprehensive briefing the night before on the sort of reception we were going to get from the media. There were going to be something like 50 journalists there and it was decided that I would respond to a lot of the questions because I spoke English.'

Walpole described how he felt as the aircraft landed at Kennedy. 'I remember talking into the little Dictaphone I was carrying and saying: 'This, surely, is one of the great moments in the history of aviation.' We taxied in and assembled in the hangar. There was a big podium and I think there were six of us there, three Frenchmen and three English guys. Arrayed in front of us was what was obviously a very hostile American news media.'

What happened next turned out to be what Walpole remembers as 'the most incredible example of the public attitude towards Concorde'. He said: 'Typical questions were: "Tell me, Captain Walpole, how long will it take you to deafen our children?" and "How long will it be before irreparable damage is done to people's houses?" That was the sort of thing we were getting.'

But then the hostile cross-examination was interrupted by the sound of the hangar doors opening. 'Everybody turned round and looked,' Walpole recalled. What they saw was Concorde being towed inside the hangar. 'The aeroplane looked magnificent. From that moment the tone of the conference changed completely. There were comments like: "What a magnificent aeroplane," and "Look at the beauty of it!" Honestly, it was sheer coincidence and not a deliberate act. I talked to the engineers afterwards and they said they had to bring it into the hangar.'

It was two decades earlier that the industry had started thinking about building a supersonic airliner. The British still believed that only a major technological leap would enable Europe to vault ahead of the US aerospace industry. In the 1950s supersonic flight was confined to a relative handful of single-seat fighters and research aircraft. The sound barrier, as it was then known, had first been pierced by Captain Chuck Yeager in the bright orange Bell X-1 in 1947. Six years later, test pilot Scott Crossfield blasted the Douglas D-558-2 Skyrocket across the sky at twice the speed of sound. Both aircraft were rocket-powered and had to be launched in-flight from carrier aircraft.

By 1953 the North American F-100 Super Sabre had become the first operational jet capable of taking off under its own power and exceeding the speed of sound in level flight. Having held the world air speed record for two brief periods, the UK seemed to have fallen behind, but in February 1956 test pilot Peter Twiss created a sensation by setting the first air speed record at more than 1,000mph (1,600kph) in the Fairey FD2 research aircraft. The new mark was left at 1,132mph (1,811kph), 300mph (1,480kph) better than the Super Sabre could manage.

To Sir George Edwards, visionary chairman of Vickers, later to become the British Aircraft Corporation, the implication was clear: it was an invitation for Britain to use this expertise to start work in earnest on a supersonic airliner. According to *The Aeroplane*, Edwards had 'publicly stated that this country would do well to take a deep breath, get out its requirements for a supersonic jet transport and get to work to build it'.

By October 1956 representatives of the major British airframe and aero-engine manufacturers were meeting at the Ministry of Aviation to discuss building a supersonic airliner. It was chaired by the ministry's permanent secretary, Sir Cyril Musgrave. Later he was to recall how the discussion had initially been about how the world's airlines were ordering US-built jet airliners. 'There was no point in developing another subsonic 'plane,' Musgrave said. 'We felt we had to go above the speed of sound or leave it.'

From that moment the die was cast. A month later the Supersonic Transport Aircraft Committee met for the first time. It was chaired by (later Sir) Morien Morgan, deputy director of the Royal Aircraft Establishment at Farnborough, who had established his reputation by leading the RAE's work on the investigation of the 1954 Comet losses. The committee's report, completed in 1959, called upon the government to proceed with the development of an SST.

As the French were thinking along similar lines – the Bristol (BAC) 233 and the Sud Aviation Super Caravelle concepts looked remarkably alike – it seemed sensible to combine resources and share the costs. And there was a political dimension. To the British government, collaboration on an advanced aircraft might help the nation's application for membership of what was then known as the Common Market.

In November 1962 the two nations sealed the treaty, which would lead them to develop a supersonic airliner as equal partners. British Aircraft Corporation and Sud Aviation would collaborate on the airframe, while Bristol Siddeley (later Rolls-Royce) and SNECMA would develop the engines. The project would be guided – and funded – by the two governments.

It was the first time in history that such a collaborative organisation had been established. Despite Britain's pioneering work on gas turbine power and its application to commercial aircraft, the two nations had accumulated less experience of high-speed flight and building jet airliners than the Americans. Veteran commentator Rene J. Francillon pointed out in 2003 that Britain and France combined had built fewer than half the number of jet airliners turned out by the US.

Developing a supersonic airliner capable of crossing the North Atlantic with up to 100 passengers in safety and with regularity was an immense challenge. The technical problems to be overcome were huge (*see* box). And such a complex aircraft required a level of testing and proving that was more extensive and thorough than any that had been applied before. Concorde was subjected to 5,000 hours of testing before it was first certified for passenger flight, making it the most tested aircraft ever.

Flight saw the technical achievement as 'difficult to exaggerate'. In January 1976, on the eve of the commencement of Concorde scheduled services, the journal noted that 'passenger supersonics, like walking on the moon and building St Peter's, approach the ultimate in human achievement'.

Yet the technical achievement was at least matched by the political and diplomatic effort involved in getting the aircraft into service. Concorde polarised opinions like few other aircraft. There were times when it seemed that the technical brilliance of the achievement was unimportant compared to its perceived danger to the planet and the human race.

Some believed that fleets of high-flying supersonic airliners would degrade the earth's ozone layer, create unacceptable levels of pollution, cause deafening sonic booms that would be heard for many miles either side of its flight path and that the roar of its engines would make the lives of people living near airports a complete misery.

Others criticised the economics of the aircraft, arguing that the cost of developing it could never be recouped and that the taxpayers of Britain and France would be saddled with an unacceptable burden. There was certainly no denying that the cost of developing Concorde was continually rising. In 1962 *Flight* put the cost at £170 million. By 1970 it had risen to an estimated £825 million. *The Economist*, never a friend of Concorde, suggested that, even before it entered service, the supersonic airliner had already cost British and French taxpayers £1.1 billion.

Concorde's operating economics represented its Achilles heel. Powered by thirsty turbojet engines at a time when most of the industry was turning to more economical turbofans meant that Concorde operations were severely hampered. Rene Francillon quoted British Airways figures showing that Concorde burned double the amount of fuel than a 747-400 to carry a quarter of its passengers.

In 1971 the USA effectively abandoned its own plans to build an SST. Two years later, Pan American and TWA announced they would not be taking up their options to buy Concordes. This news went down badly in Britain and France, some sections of the news media calling it a betrayal. The news did not help the case to operate Concordes at New York's JFK airport, which it was designed to do.

Yet North America was not among the first destinations chosen during the initial stages of the Concorde test and development programme. Only after tours of South America and the Far East and Australia, hot and high tests in Johannesburg and Madrid, and visits to West Africa and Iceland in 1971–73 did a Concorde touch down on American soil. This had been part of a deliberate policy of the manufacturers to avoid raising Concorde's profile in the USA following the cancellation of its own SST project.

On 18 September 1973, Concorde 02 arrived in Texas to help celebrate the opening of the new Dallas/Fort Worth airport. It had flown from Paris via Las Palmas and Caracas, piloted by Jean Franchi and Gilbert Defer. There were thirty-two passengers on board, one of whom was BA chairman David Nicholson. He offered a glimpse into the future when he said that by flying Concorde a business traveller could leave London early in the morning, have 'five or six hours for meetings' in New York before catching the afternoon flight home to be back at Heathrow by 22:30 hours.

The same day, Sir George Edwards was telling an audience in Washington that with Concorde, Britain and France had opened up 'an enormous lead over the USA – as big in its way as yours over us with the space programme'. He added: 'I promise you, we are not going to muff this chance. We are not going to lose our lead.'

Despite being restricted to subsonic speed, 02 still managed to lop thirty minutes off the scheduled flight time from Dallas/Fort Worth to Washington.

During its time in the USA the aircraft had been fitted into normal operating patterns and made use of existing airport equipment while keeping to a tight operating schedule. Among those impressed was Robert Hotz, editor of *Aviation Week and Space Technology,* who noted that the aircraft had flown 'airline payloads over airline stage lengths using less runway than is already available at major airports and with no smoke pollution and no higher external noise levels than many current subsonic jets'.

On 26 September 1973, at the end of its four-day visit, Concorde 02 flew home to Paris in three hours thirty-three minutes. It was the first ever North Atlantic crossing by a supersonic airliner. 'Once Concorde comes into service,' observed Joe Murphy, editor of *Air Transport World*, 'non-SST operators might just as well fold up their first-class seats and silently bow out of the premium service business on the Blue Ribband route across the North Atlantic.'

By the end of 1974 there were further Concorde visits to US cities. In June an Air France aircraft left Boston for Paris at the same time as one of the airline's Boeing 747s took off to fly in the opposite direction. By the time the two aircraft passed each other, the 747 had covered 620 miles (990km), while the Concorde had flown 2,400 (3,840). The SST landed in Paris and spent sixty-eight minutes on the ground before taking off again for Boston, where it arrived eleven minutes ahead of the 747. In September 1975 Concorde 204 (G-BOAC) made two return flights from London to Gander to become the first aircraft to make four Atlantic crossings in a single day.

In the absence of permission to operate to the USA, scheduled commercial Concorde flights began on 21 January 1976. BA inaugurated a London–Bahrain scheduled service using G-BOAA (206), while Air France launched operations to Rio de Janeiro via Dakar with F-BVFA (205). It might not have been what they had originally planned for, but it would enable Concorde's performance in scheduled service to be assessed before – hopefully – the start of services to the USA.

A fortnight later there was good news from Washington. US Transportation Secretary William Coleman approved the two airlines' request for each to operate twice daily services to New York and once a day to Washington's Dulles airport for a sixteen-month trial period. The decision had followed a major environmental study and a public hearing conducted by Coleman himself, and was made on condition that the aircraft did not exceed prescribed noise levels.

But the city of New York had other ideas. In March came the news that the Port Authority of New York and New Jersey had banned Concorde from using its airports. It was a real set-back, but hardly unexpected. New York had a history of anti-aircraft noise militancy dating back to the late 1950s. The two airlines responded by taking their case to court. *Flight* noted: 'Rarely if ever before has an airline taken a foreign airport to court over its landing rights. Thus Air France and British Airways have finally resorted to US law in their efforts to fly Concorde into the golden gate of New York.'

At issue was whether or not a local authority could flout Federal will. The air service agreement between the two countries – the Bermuda Agreement – provided for the parties to operate their chosen aircraft provided they were properly certificated. Yet the city of New York believed its ability to set airport opening hours and determine which aircraft types operated from its airports trumped Federal rulings.

By the time the issue came to court, the two airlines had the advantage of a year's Concorde operations at Dulles. They began at 13:01 hours on 24 May 1976 when British Airways Flight BA570 left Heathrow with Captain Brian Calvert, the airline's head of Concorde operations, in command. Shortly afterwards, the corresponding Air France aircraft, piloted by Captain Pierre Dudal, took off from Charles de Gaulle airport, Paris.

The captains were in touch by radio as the two supersonic airliners approached the US east coast. They slowed down to subsonic speed and by the time they reached Maryland 50 miles (80km) from Dulles the two aircraft were just 5 miles (8km) apart. They overflew the US capital before making parallel approaches to the airport, where more than 10,000 people had gathered to watch the two jets touch down, exactly on time. G-BOAC had taken three hours fifty-two minutes for the journey from London, of which two hours forty-two minutes had been flown at supersonic speed. On board were seventy-six passengers, including two British ministers and the chairman of BAC.

The Air France aircraft touched down two minutes later. It had taken three hours fifty minutes, of which two hours fifty-three minutes had been at supersonic speed. The two Concordes taxied to the front of the terminal building and parked nose-to-nose. They left one after the other on their return flights the following day. But while the Air France example used the runway designated for Concorde operations, Captain Norman Todd elected at the last moment to use the other. This was for community noise reasons but it provoked accusations that he had cheated by avoiding the noise monitors. Calvert, who was sitting in the co-pilot's seat, acknowledged later that Secretary Coleman had been embarrassed but wrote: 'No harm was meant – we were simply doing our best.'

But US air travellers were delivering their own verdicts. The services from Washington were achieving 80 per cent load factors and nearly one quarter of passengers who used them to travel to London and Paris had come from the New York area. The airlines, meanwhile, had obtained a court judgement in their favour, but that did not mean Concorde operations to JFK were about to begin. The Port Authority appealed the decision that its ban was illegal, but three appeals court judges upheld the earlier judgement. The authority's response was to ask the Supreme Court to allow the ban on Concorde operations to stand, but it refused and the way was finally cleared for the supersonic airliner to use Kennedy. After so many months of arguments in court and protests on the streets, the start of BA and Air France flights on 22 November 1977 was almost an anti-climax.

In command of the two airliners were Brian Walpole and Pierre Dudal. They flew across the Atlantic in tandem and made the expected joint approach. After landing they posed briefly together for the photographers before taxying to their own terminals. This time there was no press conference, just a few interviews, but the New York Chamber of Commerce entertained 1,000 people to lunch to mark the occasion.

After a year of monitoring, the Federal Aviation Administration reported that Concorde's noise performance at Kennedy had been in line with or slightly less than had been predicted by the 1975 Environmental Impact Statement. Yet the axe still hovered over the Concorde project. Even though BA and Air France were operating scheduled supersonic trans-Atlantic services and had carried around 300,000 passengers, after three years of service there was still uncertainty about the airliner's future.

In its early days it was under constant threat of review and even cancellation at every change of government in the UK. As MP for a Bristol constituency where Concordes were built, Tony Benn was always conscious that the project had its detractors and that some of them were inside the government itself. During his time as minister of technology Benn had been advised to pull the plug on the project by his officials. When he returned to the cabinet in 1974, Benn, as industry secretary in Harold Wilson's second administration, was 'astounded' to be told that there was now a multi-party consensus and that the project should be cancelled. But he warned his cabinet colleagues that cancelling it would mean massive compensation payments to the French. He said: 'I used to say to the Attorney General: "You're really in charge of the project because you keep warning us of the legal penalties of cancelling it."'

Then, in the 1980s, came privatisation. The Thatcher government was determined to sell shares in the nationalised British Airways as part of its flagship privatisation programme. Accordingly, in 1981 it appointed the no-nonsense Sir John King as chairman to get the carrier into shape for its public flotation.

Brian Walpole remembered the day in early 1982 when King summoned him to issue a challenge. In typically blunt fashion, King told Walpole that if the Concorde operation was not profitable by 1984 he would terminate it. 'He said, "You and I know that if BA terminates it Air France will terminate it. On the strength of that, the future of supersonic civil aviation over the next two years is effectively in your hands." I thought, what an opportunity but what a responsibility.'

Walpole believed the airline had not been making the most of its Concorde fleet. 'I'd written several papers for the board in which I said the Concorde fleet was not being properly managed,' Walpole told the author. 'I said it was a mistake that Concorde was operating under the umbrella of the 747 fleet and that it needed dedicated engineering, marketing, sales and promotions support.'

Walpole rose to King's challenge. 'We made a lot of money,' he said. 'I formed a team and we had a much closer, more intimate relationship with the engineering side because at that stage we got dedicated engineering support, as opposed to engineers who were looking after Concorde and 747s and other aircraft.'

Captain Jock Lowe worked with Walpole to put BA's supersonic operations on a sound financial footing. He retired in 2001 as manager, Concorde, and believes he was the airline's most experienced Concorde pilot. 'I flew up to seven sectors a month for nearly 25 years,' he told the author. 'That's a lot of supersonic time. But it never ceased to be fun.'

Although Concordes generally fitted in smoothly with the BA and Air France operational patterns, there were differences compared with the subsonic elements. For the flight crews these differences were apparent before each take-off. Flight planning was more complex and crews had to start work half an hour earlier than their subsonic counterparts. Fuel capacity was the critical issue and this made the weather at the other end of the journey more critical than with the other aircraft types.

'If you were at 50,000 or 60,000ft [15,240m or 18,300m] at Mach 2.0 and had an engine fail your range was dramatically cut,' Lowe told the author.

> We had a special chart which continually showed us where we could get to on three and even two engines in the event of failure. You also had to know whether you were going to turn back or keep going. If you sat and chatted about it for two or three minutes, you'd travelled another 60 miles [96km]. And that meant another 60 miles back plus the turn, which was 120 miles [192km] in diameter, so you had a lot of miles to cover while you were thinking about it.

Take-off was an especially exciting time. 'When the engine afterburners came on the acceleration was terrific,' Lowe recalled. 'You didn't see many passengers so blasé that they didn't put their newspapers down.' At 250kt the nose, drooped five degrees for take-off, was raised, together with the glass visor covering the windscreen. 'The noise halved at that point,' Lowe remembers. For every 1,000ft (300m) in height attained the power went up 2 per cent:

> You throttled right back and put the power on gradually until you were west of Reading, before full power was applied and you really started to climb. Taking off to the east towards London, we turned right at 100ft with 15 degrees of bank until reaching a certain spot over Hounslow Common. There we could put on some power and climb a bit faster. We tried to put the aeroplane over the least populated areas. In that respect we had our own routings.

These tracks took the aircraft west along the Bristol Channel, enabling the crews to go supersonic as early as possible without the risk of causing a sonic boom over land. Over the ocean the aircraft flew south of Ireland and took a great circle route that took it above the tracks allocated to subsonic aircraft. Over the ocean the aircraft flew above the jet stream.

When Concordes started operating regular trans-Atlantic flights, passengers had become accustomed to flying on wide-bodied airliners. Yet Concorde's cabin was no wider than that of a BAC One-Eleven, although for a flight of just over three hours that was hardly an issue. 'By the time we'd taken off and extinguished the seat belt signs we were starting the meals service,' Lowe recalled. With the meal finished the New York-bound aircraft would be passing Newfoundland and it was time for passengers to visit the flight deck in the days when such things were allowed.

Concorde descents started earlier than with subsonic aircraft because of the need to reduce speed well before reaching the coast. 'We had two descents,' Lowe says. 'We had the descent from supersonic height down to subsonic and then the normal one down to the ground.'

Even with the nose lowered to 12.5 degrees for landing, which happened below 270kt (310mph, 497kph), Concorde was still much faster than other aircraft. 'We used to fly at a minimum of about 250kt [288mph, 460kph] if we possibly could, only slowing down to 190 for the approach,' Lowe explains:

> We'd keep at 190kt [219mph, 350kph] to 800ft [244m] where we came down to the threshold speed, which we'd achieve by 500ft [152m]. Again, there was a bit of planning that had to go into that descent. Air traffic controllers on both sides of the Atlantic quite quickly got used to what we preferred to do and would fit us into the traffic.
>
> If we slowed right down, to say, 160kt [139mph, 223kph] the aircraft would buffet if you tried to fly level at that speed and we'd be using a tremendous amount of fuel. We didn't like doing that if we could possibly avoid it. And being a delta, Concorde was totally unstable so you had to fly it all the way down. The engineer had to call out the height: 50, 40,

> 30, 20ft [15, 12, 9, 6m]. But our rates of descent were that much higher so you had to prepare for the flare. That was no different from other large commercial aeroplanes, of course, but we were coming down a bit faster. On touchdown we were nearly 40ft (12m) above the ground, so you lost some of the normal perspectives on landing.

As the aircraft approached Heathrow airport, the air traffic controllers also had to be aware that Concorde was different. 'We had to remember that it was still doing 200mph [320kph] on the round-out,' Steve James, the airport's general manager, air traffic control said in 2003. 'There were no special procedures for handling it at the airport but it required extra thought.'

Controller Roger Clarke, who handled the last Concorde take-off from Heathrow, said: 'You had to make sure it didn't catch up the aircraft ahead. It was doing 160 knots over the threshold compared with the average of 125 knots [109mph, 174kph] for most others.'

Go-arounds were not uncommon and pilots practised them in the simulator. Lowe recalled that they were easier than with subsonic aircraft because at that stage of the flight most of the fuel had been burned off, giving the aircraft a sprightly power-to-weight ratio. The highly effective carbon-fibre brakes also gave pilots additional confidence in the event of a rejected take-off.

Many famous and influential people regularly travelled across the Atlantic on Concorde. Some were met by their personal executive jets, which had been sent on ahead earlier in the morning. Jock Lowe said he had known as many as six waiting to pick up passengers at Kennedy.

By the time British Airways and Air France celebrated ten years of Concorde operations, supersonic operations had settled down to two flights a day to New York from London and Paris. In season, BA was flying once a week to Barbados. Charters became an important part of Concorde operations and around the world journeys were especially popular. The future was looking bright.

But then this comfortable pattern was disrupted in the most tragic way possible. In the afternoon of 25 July 2000, Air France's Flight AFR 4590, operated by Concorde F-BTSC, was taking off from Paris Charles de Gaulle airport. It had been chartered on behalf of a party of holidaymakers who were to join their cruise ship at New York. Two minutes later this most iconic of aircraft had crashed, killing all 109 on board plus four on the ground. The image of the crippled Concorde trailing smoke and flames was carried on front pages around the world the next day.

The official report found that the chain of events leading to tragedy had started when a strip of metal that had fallen from the thrust reverser of the preceding aircraft on the runway had punctured one of the Concorde's main landing gear tyres. A chunk of rubber had been flung upwards, the fuel tank above had ruptured and the resulting flow of fuel ignited.

The Concorde fleets of both airlines were grounded and for the first time in twenty-four years there were no trans-Atlantic supersonic flights available. It was not until September the following year that the regulators restored the aircraft's Certificate of Airworthiness. British Airways alone had spent £17 million on safety and other modifications. Among the enhancements was a Kevlar lining to the fuel tank to prevent any recurrence of the Paris catastrophe.

Services to New York were resumed on 7 November 2001. BA chairman Lord Marshall and chief executive Rod Eddington joined guests on board G-BOAE commanded by Captain Mike Bannister and with Senior First Officer Andy Barnwell and Senior Engineer Officer Bob Woodcock also on the flight deck. In Paris, the Air France aircraft was commanded by Bannister's opposite number, Captain Edgar Chillaud. The return of the supersonic jets to New York had given the city something to smile about after the tragic events of 11 September. Passengers and crews were greeted by Mayor Rudolph Giuliani, who told them: 'This is a wonderful day.'

But less than two years later, Airbus, the European manufacturer, which by now was the holder of Concorde design authority, declared that it could no longer the support the aircraft. This left the regulators with no choice but to withdraw its Certificate of Airworthiness.

BA was the last of the two airlines to phase its supersonic jets out of service. It did so with due ceremony on 24 October 2003 when the final three Concordes arrived at Heathrow for the last time. G-BOAG, operating the final inbound scheduled flight to London from New York, was commanded by Mike Bannister.

Sitting in the glazed visual control room atop Heathrow's control tower was controller Ivor Sims. It was appropriate that he should handle that last flight; twenty-six years earlier he had handled the first scheduled Concorde flight to depart for New York.

The withdrawal from service also meant that aircrews had to be redeployed to other duties. Among them were the only two female pilots out of a total of 262 qualified to fly the supersonic jet, British Airways' Barbara Harmer and Beatrice Vialle of Air France.

Harmer made her first flight to New York as a Concorde first officer on 25 March 1993. She had left school at fifteen to become a hairdresser in her home town of Bognor Regis, West Sussex, later training as an air traffic controller, working at Gatwick Airport. She learned to fly and went on to become an instructor before becoming an airline pilot. In 1992 she was selected for Concorde training and as the first woman supersonic airliner pilot she was subjected to intense public scrutiny.

Later, Harmer achieved her command on subsonic aircraft and she retired as a Boeing 777 skipper. She died in February 2011 after losing her battle with cancer; she was fifty-seven. Air France's Beatrice Vialle made her first flight as a Concorde first officer in 2001.

Between 1976 and 2003, Concordes made nearly 50,000 flights and carried around 3.7 million passengers, 1.2 million by Air France and 2.5 million by British Airways. Will supersonic airliners again fly between Europe and North America? Only time will tell.

Creating an Icon

One of the most significant factors leading to Concorde's development was that the British Aircraft Corporation and Sud Aviation were thinking along remarkably similar lines with their concepts for a supersonic transport.

The Bristol team led by Dr Archibald Russell and Dr William Strang, and the Sud Aviation team at Toulouse under Pierre Satre and Lucien Servanty, had both envisaged a long slender fuselage with delta wings and powered by four engines mounted in pairs beneath it.

The idea of pooling resources and going for a single design was agreed in 1961 following a series of joint meetings in Toulouse and Weybridge, which led to the inter-governmental agreement signed at Lancaster House the following year.

There were differences in approach, however. At first the French were keen on a seventy- to eighty-seat aircraft for medium-haul operations, while the British favoured a 125-seat machine for trans-Atlantic flights. This was eventually to be resolved, but it did have implications for the future of the programme.

It was, though, soon apparent that the design of an airliner capable of sustained supersonic speed would require the application of fresh thinking, particularly in its aerodynamic performance. To ensure the aircraft's economic and operational viability, a balance had to be struck between handling characteristics at subsonic speeds and reducing drag during supersonic cruise. This led to the adoption of the characteristic ogival wing, which had earlier emerged from the BAC team.

Instead of bolting and riveting sections together, as had been done in the past, engineers used a process called sculpture milling. A numerically controlled milling machine carved the required shapes from solid pieces of copper-based aluminium alloy, known as RR58 in the UK and AU2GN in France. This enabled the parts to be made to closer tolerances and to be much stronger, thanks to an absence of welds or riveted joints. Even more importantly, several hundred pounds in weight were saved with no compromise in strength.

The long slender fuselage featured two-by-two seating; it was no wider than a BAC One-Eleven short-haul airliner. The long-pointed nose reduced drag, but, combined with the high angle of attack adopted by the aircraft on take-off and landing, it reduced the pilots' forward vision. A unique solution was found by adopting a drooping nose that enabled the aircraft to be configured appropriately for different stages of flight. A streamlined visor made of heat-resistant glass protected the windscreen from aerodynamic loads and high temperatures at supersonic speeds.

The air flow through the Olympus engines had to be suitable for the whole speed range. This meant that a variable geometry intake control system had to be adopted to alter the airflow available to the engines. This was achieved via complex ramp and nozzle assemblies whose main task was to maintain stability by slowing the airflow down to subsonic speeds before it then entered the engine. Concorde was also one of the first fly-by-wire aircraft. This meant that the aircraft was controlled by electrical signals sent to the hydraulically actuated flight controls.

The need to thoroughly test the aircraft not only delayed the programme, but also meant that there were two prototypes, two pre-production aircraft and two further aircraft used for development. The need for these three quite distinct versions represented part of the legacy of the initial disagreement about the aircraft's size.

Concorde 001, registered F-WTSS, was rolled out at Toulouse in December 1967 and flew for the first time in March 1969 with Andre Turcat in command. It was followed a month later by 002 G-BSST from Filton, Bristol. Brian Trubshaw was in command. By December 1975 the airworthiness authorities of both countries had awarded Concorde its Certificate of Airworthiness. The stage was set for the start of supersonic commercial operations.

British Aircraft Corporation/British Aerospace/ Aerospatiale Concorde Specification

Crew	flight crew of 3 and up to 6 cabin crew
Passengers	up to 128
Length	202ft 4in (61.66m)
Wingspan	83ft 10in (25.6m)
Height	40ft (12.2m)
Max take-off weight	408,000lb (185,070kg)
Power plant	four Rolls-Royce/SNECMA Olympus 593 turbo jets offering 38,050lb (170kN) thrust with reheat for take-off
Cruising speed	Mach 2.04 (1,350mph, 2,160kph)
Max range	4,500 miles (7,200km)
Service ceiling	60,000ft (55,385m)

Fastest Across

By the end of trans-Atlantic aviation's first century, airline travel between northern Europe and cities on the west coast of North America had settled down to journey times of six to eight hours, depending on weather and prevailing winds.

Yet between 1976 and 2003 British Airways and Air France made it possible to fly from London or Paris to New York in little more than three hours aboard their Concordes. And as they were the only supersonic airliners in operation, it was inevitable that one of them should achieve the fastest crossing by a commercial jet.

On 7 February 1996 G-BOAD flew the 3,750 miles (6,035km) from New York to London in two hours, fifty-two minutes and fifty-nine seconds. Its average speed was 1,250mph (2,010kph). The aircraft's crew, Captain Leslie Scott, First Officer Tim Orchard and Senior Engineering Officer Rick Eades, had planned to take advantage of the jet stream and had warned air traffic controllers in the US of their record-breaking intentions before taking off.

Two decades later the jet stream helped a British Airways Boeing 777-200 subsonic jet to make the same trip in five hours sixteen minutes in January 2015. The aircraft, operating Flight BA114, was reported to have reached a ground speed of 745mph (1,192kph) as it rode winds of more than 200mph (320kph) to arrive at Heathrow ninety minutes ahead of schedule.

Yet even that was not the fastest crossing recorded by a subsonic airliner. A website devoted to the type says that a British Airways BAC VC10 four-jet airliner flew from New York to Prestwick, Scotland, in five hours one minute in March 1979. The scheduled time was six hours twenty minutes. Captain Gwyn Mullet claimed to have received full ATC co-operation in beating the previous fastest crossing of five hours eight minutes made by a Boeing 707.

But all those times were put into perspective by claims made on behalf of a US Air Force Lockheed SR-71 Blackbird reconnaissance aircraft. The Mach 3 jet was timed from New York to London when it was heading for Farnborough for its first public display at the air show there in September 1974.

Majors James V. Sullivan and Noel F. Widdifield made the crossing in just under one hour fifty-five minutes. They averaged 1,807mph (2,890kph) despite having to drop down to subsonic speed to refuel from an airborne tanker. The aircraft's return to Los Angeles, flown by Captain Buck Adams and Major William Machorek, was achieved in just under three hours forty-eight minutes. It completed the 5,447-mile (8,715km) journey at an average speed of 1,436mph (2,298kph).

Chapter Eighteen

Four Engines Good, Two Engines Better

Concorde's withdrawal from service in October 2003 meant that for the first time in more than a quarter of a century there were no supersonic airliners streaking across the North Atlantic sky on their way between London and Paris and New York.

Although the airlines could not hope to recapture the prestige of supersonic operations with subsonic aircraft – including even the forthcoming Airbus A380 'Super Jumbo' – there were still opportunities for finding creative ways of maintaining the loyalty of high-yield passengers.

British Airways, which had made its Concorde fleet sweat more than Air France, came up with a way of saving time on the London–New York route while maintaining the standards of service set by Concorde. Indeed, the same iconic flight numbers were applied to the new services, which used specially configured aircraft to create the illusion of passengers having their own executive jet.

What had made it possible was another big change in the way people fly the North Atlantic and it had begun during Concorde's twenty-seven years of operation. Although it had nothing to do with speed through the air, this low-key revolution was, like all the others that had preceded it, about advancing technology.

Even since the start of regular trans-Atlantic commercial operations, the use of four engines to power the airliners in use had seemed an obvious requirement. In fact, some people joked that the only reason they flew the ocean in a four-engine aircraft was because there were none with five.

In the 1930s the US authorities introduced a rule that applied to all types of aircraft – regardless of the number of engines – and restricted operations to an area within 100 miles (160km) of a useable airport. As the aircraft of the time took around sixty minutes to cover this distance this capability was reflected in US Federal Regulation FAR 121.161, which was introduced in 1953. Generally, this applied to twin-engine aircraft – which in those days meant piston engines – and restricted operations to areas defined as being within sixty minutes' flying time of an adequate airport on one engine.

At the same time, the International Civil Aviation Organisation restricted operations to within ninety minutes of an alternate airport using all an aircraft's engines. This standard was adopted by countries other than the USA and allowed twins to be operated in south-east Asia and Australia without significant restriction.

The rule continued into the jet age, by which time it was being applied to short-haul aircraft such as the Douglas DC-9 and Boeing 737. But in 1964 the Federal Aviation Administration waived the rule in the case of three-engined airliners including the Boeing 727. This opened the way for the development of intercontinental wide-body tri-jets such as the DC-10 and TriStar. Up until the 1980s, long-range routes, especially those over water, were flown by three- and four-engine aircraft.

The first twin to challenge this established position was the technically advanced Airbus A300B4. With its innovative flight deck systems and aerodynamics, the European design was, in 1976, not only the first twin-engined wide-body airliner to enter service, but also the first to be permitted to operate across the North Atlantic under the ICAO ninety-minute rule. It was the increasing reliability of modern turbofan engines that prompted ICAO and the FAA to agree that twins could make intercontinental trans-oceanic flights provided they met certain standards of engine reliability.

The acronym ETOPS was introduced to cover such operations. At first it stood for 'extended twin operations' and later, when it covered long-range operations by all types of aircraft, for 'extended operations'. It is granted to specific aircraft based on design and testing, and approval to a specific airline to operate an ETOPS-approved aircraft is based on its pilots, maintenance, flight planning, training capabilities and history.

Boeing believed that, like the 747, its latest wide-body type, the 767, also a twin, would introduce new technical capabilities with a combination of low operating costs, long-range capability and a passenger capacity that was half that of a fully loaded 747. 'Airlines used this formula to fundamentally transform air service patterns across the North Atlantic,' Joe Sutter recalled. He wrote later:

> Before the 767, intercontinental travel relied on a skeletal route network that linked a small number of gateway hubs in each of the world's regions. Air service was via these gateways so flying from one continent to another usually required several flights. Flying to Europe from the United States, for example, mean taking a domestic flight to New York, boarding a jumbo jet for the flight across the Atlantic and then taking a third flight from the gateway airport you arrived at – generally London or Paris – to reach your final destination in Europe.

And that was the key to the 767's ability to change the way people fly. Yet during the early 1970s, when Boeing started work on the project it called the 7X7, the ability to make long over-water flights was not top of its list of attributes. Following the quadrupling of oil prices in 1973, fuel burn had become the most important selling point rather than passenger capacity. Until then economy was lower down the list of priorities.

The starting point for the 7X7 had been the airlines' need to replace their 727s and 707s on medium-range, high-density routes and, in the process, save up to 30 per cent on fuel costs. Essentially, they were looking for something bigger than the 727 but smaller than the 747, DC-10 and TriStar.

The project evolved through a number of iterations, including two- and three-engined layouts, over-wing mounted power plants and a T-tail arrangement. But when the Airbus A300 entered service in 1974 it was clear that it was a serious rival to US-built products in the medium-haul, high-density market.

By 1976 Boeing had settled on two engines to power the 7X7, which reflected growing confidence in the reliability and superior operating economics of the latest generation of powerplants. The manufacturer also planned to use new and lighter materials for the aircraft's structure. But with customer requirements for new wide-body equipment remaining somewhat uncertain, Boeing decided to hedge its bets by developing a single-aisle aircraft in parallel with the wide-body. The new project was coded 7N7. The two designs were evolved in parallel and consequently their shared design features would enable pilots to obtain a common type rating.

Although the 7X7 was America's first twin-aisle twin-jet, its fuselage was narrower than the rival Airbus A300 and A310, which offered better load-carrying ability. Boeing, however, believed that their smaller wings would restrict development potential and was determined not to fall into the same trap. With a bigger wing, the Boeing design would offer superior range and altitude performance. It would also offer greater growth potential.

On 14 July 1978 Boeing announced it was launching the 7X7 as the 767 with an order for thirty from United Airlines. According to Joe Sutter, United had exerted considerable influence over the 767's initial design. He wrote later that the airline's president, Richard Ferris, 'wanted a wide body but he wanted it to be extremely efficient. Sutter recalled:

> He didn't even want it to fly coast-to-coast non-stop because that would mean a heavier airframe and slightly less fuel efficiency. It was pretty hard to take the engines that we had and wrap an airplane around them and not exceed his objectives. He had very different thoughts about what was needed in an airliner.

Other airlines, however, told Boeing they wanted greater range. The manufacturer was sympathetic to these demands and agreed there could be no question of inhibiting the 767's performance with a wing too small for the loads it could carry. 'We believed,' said Sutter, 'that the 767 should be capable of flying all the way across the North Atlantic non-stop.' It was a critical engineering decision that represented a departure from the original concept; the 767 therefore had a larger wing than the Airbus A310 that was launched at the same time. 'We did convince United of the benefits of this large wing,' Sutter noted. 'The 767 used very little more fuel than it would have had with a small wing and it had superior take-off and landing performance.'

Boeing also worked hard to convince the FAA of the 767's suitability for such operations by collecting data from operators and analysing every engine failure and shutdown ever experienced. From this it came up with a figure of once every 50,000 years for the probability of both engines failing at the same time.

To demonstrate the 767's range capability, Boeing scheduled a 7,500-mile (12,000km) delivery flight of an Ethiopian Airlines' 767-200ER from Washington to Addis Abba on 1 June 1984. It was the subject of a special exemption by the FAA. Two months earlier, El Al had flown the first commercial trans-Atlantic 767 flight, from Montreal to Tel Aviv, in eleven hours eight minutes. The flight was operated within sixty-minute ETOPS regulations.

That spring, El Al, Air Canada and TWA had started using 767s across the Atlantic under exemptions that enabled them to fly no more than seventy-five minutes from a suitable alternate airport. The first officially approved 120-minute ETOPs operation came on 1 February 1985 when a TWA 767 operated Flight 810 non-stop from Boston to Paris. The flight took six hours thirty-two minutes. It may have opened a new era in intercontinental air travel, but, according to the journal *Aviation Week and Space Technology,* 'passengers on the first flight showed mild enthusiasm for participating in an aviation first and no concern that the aircraft was powered by two engines rather than three or four'.

Being able to observe the new rule meant that the flight had been shortened from 3,885 miles (6,216km) to 3,797 miles (6,075km). It might not have been a big distance, but when multiplied by the dozens of flights a day that would follow, the savings soon became clear. The 767 had consumed 7,000lb (3,175kg) of fuel per hour less than the Lockheed TriStar that had previously operated the service.

The CF6-80A/C2-powered 767 received 120-minute ETOPS approval three months after the P&W-equipped variants. Such was the CF6's extraordinary reliability record that this version of the 767 received 180-minute ETOPS approval in 1989, a year earlier than P&W-powered examples and three years before those using the Rolls-Royce RB211-524H.

By 1991, the number of passengers flying the Atlantic on 767s exceeded those crossing on three- and four-engined aircraft for the first time. By 2000, more than half of all crossings were being made by the 767 family of aircraft. The result, noted Joe Sutter, was that people were increasingly opting to fly direct. 'Airlines that used to funnel their domestic traffic to New York as a jump-off point for Europe found that they could fly these passengers to Europe themselves instead of giving the business away to Pan Am, TWA or European airlines at John F. Kennedy International airport.'

Captain Gil Gray, who was British Airways' chief pilot on its 767 fleet, told the author that the aircraft offered a major benefit on long, thin routes that could not support 747 operations. 'The economics of the 767 and its ability to operate over those long distances with only two engines made an enormous difference to those routes,' he said.

Initially, the 767's narrow-body sibling, the 757, received 120-minute ETOPs clearance, for charter operations. In 1988 Monarch Airlines flew the first transatlantic ETOPS operation under Civil Aviation Authority regulations. Two years later, 757s began scheduled Atlantic crossings. Like the 767, the 757 was available with extended range features, including a back-up hydraulic-powered generator and an auxiliary fan to cool the electronics. Before long, all new 757s and 767s were certificated for 180-minute ETOPs approval.

As later twins such as the Airbus A330 and Boeing 777 received 180-minute ETOPs certification as they rolled of the assembly line there was still plenty of mileage left in the 757s and 767s. The withdrawal of Concorde services left a gap in the market in the early 2000s. Clearly the subsonic jets could not match the speed, but they could offer superior comfort and convenience.

Silverjet, MaxJet Airways and Eos Airlines all represented attempts to provide such services between the UK and the US using a mix of 757s and 767s and operating from Stansted or Luton airports to New York's JFK or Newark airports. MaxJet also flew to Las Vegas and Los Angeles.

In an interview with the author in 2005, MaxJet chief executive Gary Rogliano said he was trying to create a niche in the market for his airline by filling what he called 'holes' between the offerings of other carriers. He explained how the MaxJet business philosophy had evolved:

> The original strategy was always to do point-to-point flying internationally and particularly over the Atlantic. We wanted to connect the low-cost carriers in the US with the low-cost carriers in Europe. The one thing the super-discount carriers do not have that the legacy carriers do is international connections. But the low-cost airlines don't want to fly over the oceans. It's complex and doesn't fit their business models.

So why go all-business class? 'We thought about doing a traditional two-cabin service,' Rogliano admitted, 'but we'd just be following the majors who are all recognised brand names. So we looked at an all-business class cabin and applied a low-fare model to it. I took a 767-200 with 205 seats and reduced that to 102 large comfortable seats, included meals and all amenities and found I could cut fares by 75 per cent. We followed the low-fare model but applied it to the business cabin.'

Rogliano described MaxJet's operation as essentially 'low cost but only in terms of executive travel'. This meant offering a seat with a 60in seat pitch that could be laid almost flat, quality meals and service, complementary drinks and lounges at both ends of the route. In-flight service included a four-course gourmet meal 'served on fine restaurant china with proper metal cutlery and stemmed glassware'.

Not enough people wanted to fly that way, however. MaxJet ceased operations on Christmas Eve 2007; Silverjet and Eos both hung on until the following year. But the International Airlines Group (IAG) member British Airways has been more successful. Following the sealing of a 'open skies' air services agreement between the EU and the US, which buried the Bermuda II deal that was so unpopular with American carriers, British Airways launched its own business-class trans-Atlantic operation designed to take advantage of the new accord.

Back in the 1980s, BA was one of the launch customers for the 757 and it operated a substantial fleet of the narrow-bodied airliners, but by 2008 the airline had decided that it was the ideal equipment for the trans-Atlantic operation it called Open Skies. Round trip fares between Paris and New York's JFK started at $936 and by 2012 there were three services a day. Amsterdam–New York and Paris–Washington services were quickly abandoned, but in an interview with the author, Open Skies managing director Patrick Malval insisted that his operation represented a unique product in the market.

But why should it succeed when previous attempts to offer all-business class operations between Europe and the UK had failed? Malval pointed out that load factors had consistently been above the 85 per cent mark and that operating to JFK meant support from BA as well as Oneworld partner American Airlines. But in 2017 IAG announced the termination of Open Skies and its transfer to Level, the group's long-haul, low-budget operation.

Of course, few airlines are better qualified than BA to offer special services to the high-yield passengers who used its trans-Atlantic Concorde services. In September 2009 the airline began operating a luxurious new all-business class service between London City Airport and JFK using thirty-two-seat Airbus A318 aircraft. They carried the BA001–004 flight numbers that had previously been applied to Concorde services from Heathrow.

The A318 is a member of the European manufacturer's highly successful A320 family of twin-engined airliners. It has a fuselage more than 20ft (6m) shorter than the baseline model, but its lower weight gives it an operating range 10 per cent greater. In 2006 the European Aviation Safety Agency certificated a modified control software enhancement for the A318 that enabled it to perform steep approaches at descent angles of 5.5 degrees compared to the normal 3 degrees. In May of that year the aircraft made a test flight to London City and confirmed its compatibility with the airport's limited manoeuvring and parking space.

The first of two A318s in BA colours was delivered in August 2009. Because of the weight restrictions imposed by operations from London City's short runway, westbound flights were required to call at Shannon airport on Ireland's west coast to take on more fuel. Initially, during the thirty-five-minute stopover while the aircraft was being refuelled, it was possible for passengers to pre-clear US customs and immigration and arrive at JFK as domestic passengers. Check-in times were up to forty-five minutes before flight in New York and just twenty minutes in London.

On its launch the service received high plaudits from frequent travellers. *Business Traveller* magazine's Tom Otley called it 'something totally unique'. He wrote: 'From the A318 taking off from London City and clearance at Shannon, to the onward flight to New York arriving as a domestic passenger, this is something that even jaded travellers will want to try.' The pre-clearance facility was withdrawn for BA003 in 2012 and in 2016 BA announced that it was cancelling this service as well as BA004, leaving just one round trip per day.

Meanwhile, the airlines are still being offered a choice of aircraft with two and four engines, although the future for the four-engined airliner appears to be limited. By 2018, when Boeing's advanced 787 Dreamliner twin had more than 1,280 orders and the Airbus A350 more than 850, the A380 order book seemed stuck at 317. The latest development of Boeing's 747, the -8I, had sold forty-seven.

Yet even so, the A380's first trans-Atlantic passenger service was regarded as an 'event'. On Monday, 23 November 2009 Air France launched its twice daily service between Paris Charles de Gaulle airport and New York's JFK. The double-deck airliner accommodates more than 500 passengers.

But the move to twins seems unstoppable. A new long-range 777 development that promises the ability to fly non-stop between Europe and Australia is on the way, while a variant of the latest Airbus twin, the A350, has a range of more than 9,000 miles (14,400km). Narrow-bodied types including the latest versions of the Boeing 737 and Airbus A320 and A321 operated by a new generation of budget carriers may also indicate the future development of trans-Atlantic routes.

Even though there are more than 400 non-stop routes available between Europe and North America, London–New York is the world's premier intercontinental air route and is likely to remain so. But what sort of aircraft will be flying the services in the future? Will they have blended wing bodies, electric motors or hybrid power sources? Will there be supersonic or perhaps even hypersonic aircraft? Despite the tantalising glimpses of the future revealed by the specialist press, the ideas described seem tantalisingly far off.

One thing does seem certain, though. Over the next 100 years there will still be trans-Atlantic pioneers blazing a trail for others to follow.

Boeing's Twins

The 767 was America's first wide-body twin with a fuselage width midway between the 707 and the 747 at 15ft 6in (4.77m). Although this permitted seven abreast seating with two parallel aisles for the whole length, the 767 was narrower than other wide-body aircraft and produced less drag.

This, however, mean that its cargo hold was not wide enough for two standard LD3 unit load devices side-by-side, so a smaller container, the LD2, had to be produced specially for the 767. Boeing's engineers used computer-aided design techniques for up to 40 per cent of the design drawings and conducted 26,000 hours of wind tunnel testing.

The 767's key feature, however, was its wing design. An aft-loaded profile was proposed to distribute lift more evenly across the surface; the wings were also large in relation to the fuselage and this was to give the 767 its superior high-altitude cruise performance. It also allowed for enhanced fuel capacity and scope for bigger variants.

Three basic 767 variants were planned. The baseline 767-100 would feature 180 seats – but was never produced as it was too close to the 757 in capacity – while the -200 would accommodate 210. Customers were offered a choice between the Pratt & Whitney JT9D and the General Electric CF6, both of which generated about 48,000lb of thrust. Later, the Rolls-Royce RB211 would be an option.

Construction of the prototype began in July 1979. It was rolled out on 4 August 1981 displaying the registration N767BA. Boeing test pilots Tommy Edmonds, Lew Wallick and John Brit were on board for the 26 September maiden flight, which was uneventful except that the undercarriage could not be retracted due to a hydraulic leak. By this time, Boeing had logged 173 orders from seventeen customers including Air Canada, All Nippon Airways, Britannia Airways and TWA.

Six aircraft, four with JT9D engines, participated in the ten-month flight testing programme. Test pilots noted that the 767 was generally easy to fly and more manoeuvrable than other wide-bodied aircraft. After 1,600 hours of testing the Pratt & Whitney-powered 767-200 was certificated by the US Federal Aviation Administration and Britain's Civil Aviation Authority.

United received its first example in August 1982 and put the 767 into service on its Chicago–Denver route a month later. Delta took delivery of its first CF6-powered aircraft in October. One of the first things the airlines noticed about the 767 was its reliability. In its first year of operations the 767 logged an amazing 96 per cent dispatch reliability, which exceeded the industry average for new aircraft. The airlines also liked its economic performance, while passengers commented favourably on its comfort.

Boeing wasted little time in capitalising on the 767's growth potential. The extended range 767-200ER was launched in December 1983 and offered greater fuel capacity along with the ability to carry heavier payloads up to 5,700 miles (9,120km).

ETOPS approval opened the door to more direct and fuel efficient trans-Atlantic routings, as well as north Pacific routes to Asia. 'The 767 led the way in market fragmentation during the mid-1980s by allowing passengers in smaller US cities to fly directly to European cities,' said Joe Quinlivan, Boeing's 767 Programme General Manager. 'These point-to-point flights have become the model for new market development.'

Britannia Airways was the first airline in Europe to operate the 767, as it had been with the Boeing 737. It ordered two 767-200s in 1980 with options on a further two, and the first arrived in 1984. By the end of the decade, further orders had taken its commitment to eight in service with eight on order.

The first major change to the 767 came with the -300. Launched in February 1983, it featured a 21ft (6.4m) fuselage stretch. It also had a strengthened undercarriage and wing structure, and it flew for the first time in January 1986. Japan Airlines became the first customer that September.

The -300ER (Extended Range) variant had a higher gross weight and development began in January 1985. The first customer was American Airlines, which ordered fifteen in March 1987 and took delivery of its first example the following February. A freighter version with strengthened floor and undercarriage was developed on the basis of an order for up to sixty from United Parcel Service. Deliveries began in 1995.

British Airways took delivery of its first Rolls-Royce RB211-524H-powered 767-300ER in February 1990. It had placed its first order for eleven in 1987 and eventually had twenty-eight, the last six being delivered in 1991.

Captain Gil Gray was Chief Pilot on BA's 757 fleet at the time and he 'lobbied hard' for the 767. 'I made my feelings well known that it would be sensible to buy the 767 as opposed to one or two of the others which were on offer simply because we had a large body of pilots who would be able to fly it straight away.'

Captain Gray recalled the 767 as:

> a lovely aeroplane to operate, comfortable and with all the gizmos. To my mind it was well designed for the job it had to do; it was fun to fly and it has proved to have been a sound design. Its operating economics were, for the time, as good as any. That's shown by the sales figures and the longevity.

In the mid-1990s Boeing began considering a further stretch, taking the fuselage length to more than 200ft, in the -400ER. Featuring up to 15per cent more capacity than the -300, it could carry up to 373 passengers. The aircraft introduced raked wing tip extensions and an interior layout similar to that of the 777. Delta ordered twenty-one and Continental followed with an order for twenty-six.

The first 767-400 was rolled out at Everett in August 1999 and made its first flight two months later. Continental received its first examples in July 2000 and put the type into service in September. A further variant, the -400ERX with still more range, was ordered by Kenya Airways in September 2000. These were switched to -200ERs the following year when the project was cancelled.

The 1,000th 767, a -300ER for All-Nippon Airways, was rolled out on 2 February 2011. It was the last to be built on the original 767 assembly line. From then on production was shifted to a new and smaller location at the Everett plant to make way for the 787 Dreamliner. By 2016 767s were still operating 19 per cent of twin-engined services across the North Atlantic, not far behind the Airbus A330 and the Boeing 777.

Work on the 757 proceeded concurrently with the 767, although development of the bigger aircraft was several months further advanced. To reduce risk, Boeing had combined design work on the two types, resulting in similar features such as handling characteristics and interior fittings. They also shared avionics, flight management systems and instrumentation.

British Airways and Eastern Airlines were the launch customers and both took delivery during 1983, although the US carrier was the first to put the aircraft into service. Both selected the Rolls-Royce RB211-535C turbofan to power their aircraft, but the Pratt & Whitney PW2000 was also available.

In 1996 Boeing launched the stretched 757-300, which could seat up to 280 passengers and had about 10 per cent lower seat-mile operating costs than the -200. The first -300 was delivered in 1999.

In late 2003 Boeing decided to end 757 production because it considered the increased capabilities of the newest 737s as well as the new 787 fulfilled the 757 market's needs. Accordingly, on 28 November the 757's twenty-three-year production run was ended when the 1,050th example was delivered to Shanghai Airlines.

Specifications

Boeing 767-200

Flight crew	2
Passengers	up to 245, one class
Length	159ft 2in (48.5m)
Wingspan	156ft 1in (47.6m)
Height	52ft (15.9m)
Maximum take-off weight	395,000lb (179,169kg)
Power plant	two Pratt & Whitney JT9D-7R4D turbofans each developing 48,000lb (213.5kN) lb thrust; two Pratt & Whitney PW4056 turbofans each developing 56,750lb (252.4kN) thrust or two General Electric CF6-80A2 turbofans each developing 48,670lb (216.5kN) thrust
Cruising speed	528mph (845kph)
Range	5,841 miles (9,401km)
Service ceiling	43,100ft (13,137m)

Boeing 757-200

Flight crew	2
Passengers	up to 228, one class
Length	155ft 3in (47.3m)
Wingspan	124ft 10in (38m)
Height	44ft 6in (13.6m)
Maximum take-off weight	255,000lb (115,660kg)
Power plant	two Rolls-Royce RB211-535E4 turbofans each developing up to 43,500lb (193kN) thrust or two Pratt & Whitney PW 2000-43 turbofans each developing up to 42,600lb (189kN) thrust
Cruising speed	531mph (918kph)
Range	4,505 miles (7,250km)
Service ceiling	42,000ft (13,000m)

Appendix

Replica Vimys

Three full-size replicas of the Vickers Vimy have been built in addition to a cockpit section completed by the manufacturer for the London Science Museum in the early 1920s.

British Lion Films commissioned a replica for a film about Alcock and Brown's trans-Atlantic flight. Although the film was never made, the model was completed and went on static display at the Battle of Britain air display at RAF Biggin Hill in 1955. Its subsequent fate remains a mystery, although there were reports that it was dismantled and stored in east London until at least the late 1980s.

In 1969 an airworthy Vimy replica was built by the Vintage Aircraft Flying Association at Brooklands. Registered G-AWAU, it was first flown on 3 June by British Aircraft Corporation test pilot D.G. 'Dizzy' Addicott and Peter Hoare. It was powered by genuine 375hp Rolls-Royce Eagles similar to those used on the original Vimys. They were rebuilt by Rolls-Royce staff from original engines and parts.

The aircraft was badly damaged by fire that summer. Until February 2014 it was on display at the RAF Museum at Hendon and is currently stored in dismantled condition at the Museum's Stafford facility.

A second Vimy replica is on display at the Brooklands Museum in Surrey on the site of the Vickers factory where the original was built. It was commissioned by Peter McMillan and built in California by John La Noue in 1994 to re-enact the Vimy's three long-distance flights of 1919–20. Registered NX71MY, it first flew on 30 July 1994. Later that year McMillan and Lang Kidby piloted it to Australia to commemorate the seventy-fifth anniversary of Ross and Keith Smith's flight.

In 1999 Mark Rebholz and La Noue flew it to South Africa and on 2–3 July 2005 it achieved the ultimate goal when Steve Fossett and Mark Rebholz re-enacted Alcock and Brown's flight from St John's, Newfoundland, to Clifden, Ireland. They made the crossing in just under nineteen hours.

In 2006 ownership passed to the American ISTAT Foundation and the aircraft was maintained in airworthy condition at Dunsfold Park by Brooklands Museum volunteers. It was donated to Brooklands Museum Trust on 26 August 2006. It made subsequent appearances at the Farnborough air show and the Goodwood Revival, and the aircraft was at the 2009 Connemara air show in Ireland to commemorate the ninetieth Anniversary of Alcock and Brown's flight.

It was then decided to retire the aircraft from flying. The final flight was made by John Dodd, Clive Edwards and Peter McMillan from Dunsfold to the grass landing strip at Mercedes-Benz World adjacent to the Brooklands Museum on 15 November 2009.

Four days later it was dismantled and moved to the museum, where it was reassembled inside the main hangar by a team of volunteers. Two days later a special Brooklands Vimy Exhibition was officially opened by Peter McMillan. This unique aircraft remains on public display at the museum.

Vickers Vimy Replica Specification

Wingspan	68ft (20.75m)
Length	43ft 6in (13.25m)
Height	16ft 4in (4.98m)
Cruising speed	75mph; stalling speed 40mph
Empty weight	7,642lb (3,467kg), maximum gross weight: 12,500lb (5,445kg)
Power plant	Australia flight – two 7.4lit Chevrolet V8s in NSCAR racing trim; South Africa flight – two 5.4lit BMW M73 V12s producing 321bhp (240kw), Atlantic flight: two 8.1lit Orenda OE600 V8s producing 600bhp (450kW) maximum and 500bhp (375kW) continuous, driving 10ft 6in (3.20m) diameter, four-bladed Műhlbauer propellers
Fuel capacity	530gal (2,400lit); 40 gallons/hour average burn yields approximately 15 hours' endurance
Ceiling	13,800ft (4,200m) at 9,000lb; 1,200ft (365m) at 12,500lb

Bibliography

Published Books

Alcock, Sir John and Whitten Brown, Sir Arthur, *Our Transatlantic Flight*, William Kimber, London, 1969

Andrews, C.F. and Morgan, E.B., *Vickers Aircraft Since 1908*, Putnam, 1969/1988

Anon, *Atlantic Bridge, The Official Account of RAF Transport Command's Ocean Ferry*, HMSO, 1945

Beaty, Betty Campbell, *Winged Life*, Airlife, 2001

Beaty, David, *The Water Jump: The Story of Transatlantic Flight*, HarperCollins, 1977

Becher, Thomas, *Boeing 757 and 767*, Crowood Press, 1999

Botting, Douglas, *Dr Eckener's Dream Machine*, HarperCollins, 2002

Bowman, Martin W., *Boeing 747*, Crowood Press, 2000

Calvert, Brian, *Flying Concorde, the Full Story*, Airlife, 2002

Christie, Dr Carl A., *Ocean Bridge: The History of RAF Ferry Command*, University of Toronto Press, 1995

Colston-Shepherd, Edwin, *Great Flights*, Adam and Charles Black, London, 1939

Corrigan, Douglas, *That's my Story*, E.P. Dutton, New York, 1938

Davies, R.E.G., *A History of the World's Airlines,* Oxford University Press, 1964

Davies, R.E.G., *Lindbergh, the Man and his Aircraft,* Paladwr Press, 1997

Davies, R.E.G., *Comet,* Paladwr Press, 1999

Davies, R.E.G., *Rebels and Reformers of the Airways*, Airlife, 1987

Gardner, Charles (ed) *Fifty Years of Brooklands,* Heineman, 1956

Eglin, Roger and Ritchie, Berry, *Fly My, I'm Freddie!*, Weidenfeld and Nicholson, London, 1980

Hales-Dutton, Bruce, *Comet: Unseen Images from the Archives*, Danann Publishing, 2014

Hawker, H.A. and Mackenzie Grieve, K., *Our Atlantic Attempt*, Methuen & Co, London, 1919

Keil, Charles, *From the Cockpit 15, Sabre,* Ad Hoc Publications, 2011
Lindbergh, Charles, *Spirit of St Louis*, Scribner, 2003
Lynch, Brendan, *Yesterday We Were in America*, Haynes Publishing, 2012
Norris, Geoffrey, *The Short Empire Boats*, Profile Publications, 1966
Orlebar, Christopher, *The Concorde Story*, Osprey Publishing, 2004
Post, Wiley and Gatty, Harold, *Around the World in Eight Days, The Flight of the Winnie Mae*, John Hamilton, London
Powell, Air Commodore Griffith, *Ferryman, From Ferry Command to Silver City,* Airlife Publishing, 1982
Sampson, Anthony, *Empires of the Sky*, Hodder & Stoughton, 1984
Simons, Graham M, *Concorde Conspiracy – the Battle for American Skies, 1962–77*, The History Press, 2012
Simpson, Richard V, *The US Navy-Curtiss Flying Boat NC-4*, Fonthill Media, 2016
St John Turner, Paul, *The Vickers Vimy*, Patrick Stephens Ltd, 1969
Staniland, Prof Martin, *Government Birds, Air Transport and the State in Western Europe*, Rowman and Littlefield, 2003
Sutter, Joe, *747, Creating the World's First Jumbo Jet and Other Adventures from a Life in Aviation,* Smithsonian Books, Washington DC, 2006
Tanaka, Shelley, *The Disaster of the Hindenburg*, Madison Press, Toronto 1993
Woodley, Charles, *Bristol Britannia*, The Crowood Press, 2002

Newspapers and Magazines

The Daily Mail
The Times
The New York Times
Aviation Week and Space Technology
The Aeroplane
The Aviation Historian
Flight
Airport
Aviation News

Websites

www.FlightGlobalarchive.com
www.airships.net
www.concordesst.com

Index

C